Antiphospholipid Antibodies and Syndrome

Special Issue Editor
Ricard Cervera

MDPI • Basel • Beijing • Wuhan • Barcelona • Belgrade

MDPI

Special Issue Editor
Ricard Cervera
Department of Autoimmune Diseases, Institut Clínic of Medicine and Dermatology Hospital Clínic
Spain

Editorial Office
MDPI
St. Alban-Anlage 66
Basel, Switzerland

This edition is a reprint of the Special Issue published online in the open access journal *Antibodies* (ISSN 2073-4468) from 2016–2017 (available at: http://www.mdpi.com/journal/antibodies/special issues/antibody antiphospholipid syndrome).

For citation purposes, cite each article independently as indicated on the article page online and as indicated below:

Lastname, F.M.; Lastname, F.M. Article title. *Journal Name* **Year**, *Article number*, page range.

First Editon 2018

ISBN 978-3-03842-947-0 (Pbk)
ISBN 978-3-03842-948-7 (PDF)

Table of Contents

About the Special Issue Editor v

Preface to "Antiphospholipid Antibodies and Syndrome" vii

Yik C. Ho, Kiran D. K. Ahuja, Heinrich Körner and Murray J. Adams
β_2GP1, Anti-β_2GP1 Antibodies and Platelets: Key Players in the Antiphospholipid Syndrome
doi: 10.3390/antib5020012 . . . . **1**

Gabriella Lakos, Chelsea Bentow and Michael Mahler
A Clinical Approach for Defining the Threshold between Low and Medium Anti-Cardiolipin
Antibody Levels for QUANTA Flash Assays
doi: 10.3390/antib5020014 . . . . **17**

Karl J. Lackner and Nadine Müller-Calleja
Antiphospholipid Antibodies: Their Origin and Development
doi: 10.3390/antib5020015 . . . . **25**

Anna Brusch
The Significance of Anti-Beta-2-Glycoprotein I Antibodies in Antiphospholipid Syndrome
doi: 10.3390/antib5020016 . . . . **35**

José A. Martínez-Flores, Manuel Serrano, Jose M. Morales and Antonio Serrano
Antiphospholipid Syndrome and Kidney Involvement: New Insights
doi: 10.3390/antib5030017 . . . . **46**

José A. Gómez-Puerta, Gerard Espinosa and Ricard Cervera
Antiphospholipid Antibodies: From General Concepts to Its Relation with Malignancies
doi: 10.3390/antib5030018 . . . . **58**

Gary W. Moore
Current Controversies in Lupus Anticoagulant Detection
doi: 10.3390/antib5040022 . . . . **67**

Jessica Bravo-Barrera, Maria Kourilovitch and Claudio Galarza-Maldonado
Neutrophil Extracellular Traps, Antiphospholipid Antibodies and Treatment
doi: 10.3390/antib6010004 . . . . **82**

Alexey Kolyada, David A. Barrios and Natalia Beglova
Dimerized Domain V of Beta2-Glycoprotein I Is Sufficient to Upregulate Procoagulant
Activity in PMA-Treated U937 Monocytes and Require Intact Residues in Two Phospholipid-
Binding Loops
doi: 10.3390/antib6020008 . . . . **94**

Chau-Ching Liu, Travis Schofield, Amy Tang, Susan Manzi and Joseph M. Ahearn
Potential Roles of Antiphospholipid Antibodies in Generating Platelet-C4d in Systemic
Lupus Erythematosus
doi: 10.3390/antib6030009 . . . . **110**

About the Special Issue Editor

Ricard Cervera, MD, PhD, FRCP, is Senior Consultant and Head of the Department of Autoimmune Diseases, Hospital Clínic, Barcelona, Director of the Research Team on Systemic Autoimmune Diseases at the Institut d'Investigacions Biomèdiques August Pi i Sunyer (IDIBAPS) of Barcelona, Professor at the Department of Medicine and Coordinator of the Master in Autoimmune Diseases at the Universitat de Barcelona, Barcelona, Catalonia, Spain. Dr. Cervera's current major research interests include the clinical and epidemiological aspects of systemic autoimmune diseases, particularly systemic lupus erythematosus and the antiphospholipid syndrome (APS), with special focus on its "catastrophic" variant, for which a patient registry ("CAPS Registry") was created and is coordinated by Dr. Cervera. Dr. Cervera has presented over 400 invited lectures and has published more than 800 scientific papers (h index: 67). His academic activities include invited Professorships in several European and Latin American Universities. He is co-editor of 30 books, Associate Editor of "Lupus Science & Medicine", and member of the Editorial Board of 20 medical journals.

Preface to "Antiphospholipid Antibodies and Syndrome"

The antiphospholipid syndrome (APS) is defined by the occurrence of often multiple venous and arterial thromboses and pregnancy morbidity (abortions, fetal deaths, premature births), in the presence of antiphospholipid antibodies, namely, lupus anticoagulant, anticardiolipin antibodies, or anti-β2 glycoprotein-I antibodies. The APS can be found in patients having neither clinical nor laboratory evidence of another definable condition (primary APS) or it may be associated with other diseases, mainly systemic lupus erythematosus (SLE), and, occasionally, other autoimmune conditions, infections, and malignancies. Rapid chronological occlusive events, occurring over days to weeks, have been termed catastrophic APS. Other postulated APS subsets include microangiopathic (4) and seronegative APS.

In the 35 years since the original description of this syndrome, advances in the recognition of both the clinical and the underlying aspects of the condition have been notable. Single vessel or multiple vascular occlusions may give rise to a wide variety of manifestations in the APS. Any combination of vascular occlusive events may occur in the same individual, and the time interval between them varies considerably from weeks to months or even years. As a consequence, the APS is, at last, gaining recognition in all branches of medicine, from obstetrics to cardiology, from psychiatry to orthopedics. This volume highlights several current concepts on the pathogenesis, diagnosis, clinical manifestations, and therapy of APS. It also brings together many of the internationally known experts in this field. Although the APS is a relatively "young" syndrome, it seems to "replace" SLE in its diversity of manifestations, and, accordingly, the number of clinical and scientific publications and of medical meetings on APS is growing.

Ricard Cervera
Special Issue Editor

antibodies

MDPI

Review

β2GP1, Anti-β2GP1 Antibodies and Platelets: Key Players in the Antiphospholipid Syndrome

Yik C. Ho [1], Kiran D. K. Ahuja [1], Heinrich Körner [2] and Murray J. Adams [1,*]

[1] School of Health Sciences, University of Tasmania, Locked Bag 1322, Launceston, Tasmania 7250, Australia; yik.ho@utas.edu.au (Y.C.H.); kiran.ahuja@utas.edu.au (K.D.K.A.)
[2] Menzies Institute for Medical Research, University of Tasmania, Private Bag 23, Hobart, Tasmania 7001, Australia; heinrich.korner@utas.edu.au
[*] Correspondence: murray.adams@utas.edu.au; Tel.: +61-3-6324-5483

Academic Editor: Ricard Cervera
Received: 5 April 2016; Accepted: 26 April 2016; Published: 6 May 2016

Abstract: Anti-beta 2 glycoprotein 1 (anti-β2GP1) antibodies are commonly found in patients with autoimmune diseases such as the antiphospholipid syndrome (APS) and systemic lupus erythematosus (SLE). Their presence is highly associated with increased risk of vascular thrombosis and/or recurrent pregnancy-related complications. Although they are a subtype of anti-phospholipid (APL) antibody, anti-β2GP1 antibodies form complexes with β2GP1 before binding to different receptors associated with anionic phospholipids on structures such as platelets and endothelial cells. β2GP1 consists of five short consensus repeat termed "sushi" domains. It has three interchangeable conformations with a cryptic epitope at domain 1 within the molecule. Anti-β2GP1 antibodies against this cryptic epitope are referred to as 'type A' antibodies, and have been suggested to be more strongly associated with both vascular and obstetric complications. In contrast, 'type B' antibodies, directed against other domains of β2GP1, are more likely to be benign antibodies found in asymptomatic patients and healthy individuals. Although the interactions between anti-β2GP1 antibodies, β2GP1, and platelets have been investigated, the actual targeted metabolic pathway(s) and/or receptor(s) involved remain to be clearly elucidated. This review will discuss the current understanding of the interaction between anti-β2GP1 antibodies and β2GP1, with platelet receptors and associated signalling pathways.

Keywords: anti-beta 2 glycoprotein 1 antibodies; beta 2 glycoprotein 1; platelet; antiphospholipid antibody; antiphospholipid syndrome; systemic lupus erythematosus

1. Introduction

Anti-phospholipid (APL) antibodies are a heterogeneous group of autoantibodies targeting different phospholipid binding protein antigens. These autoantibodies include lupus anti-coagulant (LAC), anti-cardiolipin (aCL), anti-beta 2 glycoprotein 1 (anti-β2GP1), and anti-prothrombin antibodies [1]. APL antibodies dysregulate normal cellular activities and are associated with recurrent thrombosis (venous, arterial, and microvascular), pregnancy complications (e.g., obstetric failure, pre-eclampsia and eclampsia), and non-specific manifestations (e.g., thrombocytopenia, heart valve disease, chorea, livedo reticularis/racemosa, and nephropathy) [2]. APL antibodies are also present in 1%–5% of healthy populations, including children [3,4]. These populations appear to be asymptomatic, since their autoantibodies are associated with low reactivity [4].

Persistently high levels of APL antibodies, together with specific clinical manifestations, are required for the diagnosis of antiphospholipid syndrome (APS) [1]. APS can occur in isolation or in association with underlying autoimmune diseases such as systemic lupus erythematosus (SLE). The Sydney criteria for the diagnosis of APS recommend that three standard diagnostic assays are used to detect APL antibodies [5]. These diagnostic assays include two enzyme-linked immunosorbent assays (ELISA) that directly detect APL antibodies binding to cardiolipin-β_2GP1 complexes, or β_2GP1 only. The third is a clotting assay which indirectly detects APL antibodies by measuring their functional effects on the coagulation system (LAC activity, Table 1) [1,3,6]. Although these assays detect overlapping subpopulations of autoantibodies, their correlation with the clinical manifestations of APS can be varied. LAC assays are superior for detecting pathological subpopulations of APL antibodies when the quality of plasma is maintained [7]. ELISAs for aCL and anti-β_2GP1 antibodies, however, are weakly associated with thrombotic complications. This may be due to poor standardisation of assays, variable sources and the integrity of β_2GPI, the secondary calibration process, and/or the assessment and derivation of cut-off values [8]. Consequently, a combination of these tests is used to determine the clinical risk. Patients with persistently high APL antibodies titres (positive in ELISA) and positive LAC activities on at least two occasions, 12 weeks apart, are at higher risk of thrombosis and/or pregnancy complications [1].

The criteria for the diagnosis of APS are well established, yet the interactions between APL antibodies, targeted antigens, and receptors remain unclear. Anti-β_2GP1 antibodies and their target, β_2GP1, have become a focus of research for their potential role in thrombosis and pregnancy complications [9]. β_2GP1-dependent LAC antibodies demonstrate a stronger correlation with thrombosis compared to β_2GP1-independent LAC antibodies [10,11]. Similarly, β_2GP1-dependent aCL antibodies are more highly associated with APL antibodies-related complications compared to transient β_2GP1-independent aCL antibodies induced by infections [12]. Many potential mechanisms of interaction between anti-β_2GP1 antibodies, β_2GP1, and cells—e.g., platelets, endothelial cells and monocytes—have been suggested [13]. However, studies investigating the effects of anti-β_2GP1 antibodies and β_2GP1 on platelets [14–16] may help lead to an improved understanding of their interactions, and consequently, their impact on the haemostatic system [17]. Activation of platelet receptor(s)/metabolic pathway(s) by anti-β_2GP1 antibodies and β_2GP1 may result in excessive clot formation and potentially initiate thrombosis and/or pregnancy complications [14–16]. Therefore, this review discusses the current understanding of the characteristics and interactions between β_2GP1 and anti-β_2GP1 antibodies in relation to platelet receptors and function.

Table 1. Detection of anti-phospholipid antibodies and their clinical significance.

Assays	Principle of Detection	Antibodies Detected	Clinical Significance [5]
LAC	Clotting assay	LAC (mainly against β_2GP1 and prothrombin)	• Strong correlation with thrombosis [18] and pregnancy morbidity [19]. • Weak correlation with thrombosis and pregnancy morbidity [5,20].
aCL antibody	Immunological assay	aCL antibody (IgG, IgM, IgA)	• Possible false positive in IgM assay caused by rheumatoid factor or cryoglobulins [21,22]. • IgA assay only useful to identify patient subgroups with specific clinical manifestations [5].
Anti-β_2GP1 antibody	Immunological assay	Anti-β_2GP1 antibody (IgG, IgM, IgA)	• Independent risk factor for thrombosis [23] and pregnancy complications [24]. • Higher specificity and lower inter-laboratory variation compared to aCL assay [5]. • Clarifies pre-eclampsia and/or eclampsia in pregnant women with negative aCL [24]. • Possible false positive in IgM assay caused by rheumatoid factor or cryoglobulins [5]. • Presence of IgA might not associate with any clinical manifestation [5].

Table 1. *Cont.*

Assays	Principle of Detection	Antibodies Detected	Clinical Significance [5]
Anti-prothrombin antibody	Immunological assay	Anti-prothrombin and anti-phosphatidylserine-prothrombin complex	• May serve as a confirmatory assay for LAC [25]. • Association with thrombotic risk still needs to be clarified [5].

Information collated from Miyakis *et al.* (2006) [5]. Abbreviations: LAC, lupus anti-coagulant; aCL antibody, anti-cardiolipin antibody; Ig, Immunoglobulin; anti-β_2GP1 antibody, anti-beta 2 glycoprotein 1 antibody.

2. β_2GP1

APL antibodies were originally thought to bind directly to phospholipids [26]. In the 1990s, three independent groups demonstrated that APL antibodies actually interacted with phospholipids via β_2GP1 [27–29], significantly raising the interest in this protein. β_2GP1 had been discovered earlier in 1961 [30], and its amino acid sequence determined in 1984 [31]. It was misnamed apolipoprotein H [32], since it is not an integral part of lipoproteins. Once synthesised in the liver and placenta, β_2GP1 circulates in blood at a concentration of approximately 4–5 μM. Blood levels of β_2GP1 are higher in older individuals and in patients with APS, but are lower in pregnant women and patients with stroke and myocardial infarction [33].

β_2GP1 is an evolutionarily conserved single chain anionic phospholipid-binding glycoprotein, with a molecular weight of approximately 43 kDa [34–36]. It belongs to the complement control protein superfamily [37] and consists of 326 amino acids that are arranged in five short consensus repeat, termed "sushi" domains [31,38,39]. The first four domains, each comprising approximately 60 amino acids, are conserved sequences linked together by two disulfide bridges. The fifth domain (DV), however, is a modified form with 82 amino acids. It contains a six residue insertion, a 19-amino acid C-terminal extension and an additional disulfide bond that includes a C-terminal cysteine. These positively charged lysine-rich amino acids (282–287) determine the affinity of β_2GP1 for anionic phospholipids and negatively charged molecules. DV also adopts a flexible hydrophobic loop (amino acids 311–317), containing a Trp-Lys sequence which is potentially able to insert into membranes. β_2GP1 has four N-glycosylation sites (Arg143, Arg 164, Arg 174, and Arg 234) located in third domain (DIII) and fourth domain (DIV). There is also one O-linked sugar on Thr130 in β_2GP1 that accounts for approximately 20% *w/w* of the total molecular mass [40].

2.1. Conformations of β_2GP1

β_2GP1 adopts many post-translational modifications which alter the structure and function of the molecule and the exposure of the cryptic epitope [41]. Among them, three interchangeable conformations are more commonly reported (Figure 1). The first conformation was reported by two groups [38,42] based on the crystal structure of the protein. In this conformation, first four domains are stretched with DV at a right angle to the other domains, resembling a J-shape, fish-hook or 'hockey stick' conformation. The second reported conformation is S-shaped, as demonstrated using small-angle X-ray scattering [43]. This conformation contains carbohydrate chains from DIII–IV that are twisted and positioned on DI. The third conformation is a common 'closed' circular formation present in plasma where DI interacts with DV. This circular formation was initially proposed by Koike *et al.* in 1998 [44], and later directly visualised by Agar *et al.* (2010) using electron microscopy [41].

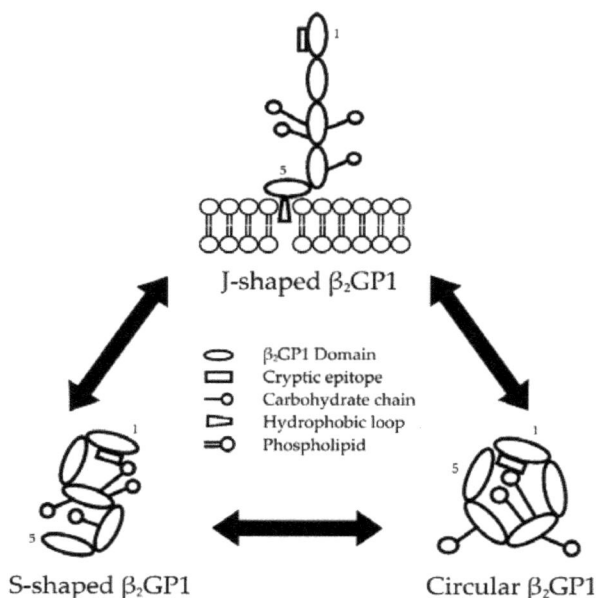

Figure 1. The interchangeable conformations of beta-2-glycoprotein 1 (β_2GP1). β_2GP1 is able to transform between three conformations: J-shaped, S-shaped, and circular β_2GP1. Cryptic epitopes in S-shaped are shielded by carbohydrate chains [43]. Whereas, cryptic epitopes in circular β_2GP1 are shielded by both carbohydrate chains and domain V [41,44]. Binding of domain V positively charged amino acids and hydrophobic loop to phospholipid membrane breaks the shield on domain I [41]. This exposes the cryptic epitope and allows the binding of clinically significant anti-domain-I-β_2GP1 antibody.

2.1.1. Transformation between β_2GP1 Conformations

The discovery of three interchangeable β_2GP1 structures led to increased understanding of the interaction between anti-β_2GP1 antibodies and β_2GP1. These conformational alterations determine the exposure of the cryptic epitope which includes arginine 39–arginine 43 (R39–R43), DI–II interlinker, and possibly aspartic acid residues at positions 8 and 9 [45]. Anti-domain-I-β_2GP1 (anti-DI-β_2GP1) antibodies targeting this discontinuous epitope are highly associated with APL antibodies-related clinical manifestations [46,47].

β_2GP1 is suggested to circulate in an S-shaped or a circular conformation, with less than 0.1% of β_2GP1 in circulation present in the J-shaped conformation [41,47]. The cryptic epitope in both S-shaped and circular β_2GP1 is shielded by carbohydrate chains positioned on top of DI [43,48]. In circular β_2GP1, these negatively-charged carbohydrate chains are also proposed to neutralise the positively-charged DI, allowing the binding of DV [47]. Therefore, S-shaped β_2GP1 may represent an intermediate form of the molecule as it transforms from a circular to J-shaped conformation [47]. When positively charged amino acids and hydrophobic loop in DV interact with anionic surfaces, β_2GP1 opens out to the J-shaped conformation, breaking the shield on DI and exposing the cryptic epitope [41].

2.1.2. Factors Affecting β_2GP1 Conformation

The conformation of β_2GP1 is dependent on its interaction with anionic surfaces. Its affinity decreases in the presence of ethylene-diamine-tetra-acetic acid (EDTA) [49], and high concentrations of bivalent cations—e.g., calcium and magnesium ions [50]. β_2GP1 that has been cleaved at DV is also known to have lower affinity [51]. Conversely, dimerisation [52] and increasing β_2GP1 concentration [50] elevate its affinity. Besides exposure to anionic surfaces, alternations to pH and salt concentration *in vitro* allow structural transformation of β_2GP1 [41]. High pH and salt concentrations convert circular β_2GP1 into the J-shaped conformation, and *vice versa* at a low pH and salt concentration. It has also been speculated that these alterations in pH and salt concentration possibly affect the hydrophilic interaction that may be present between DI and DV [41].

APS patients have been proposed to have higher oxidative stress compared to healthy individuals [53]. Oxidative stress favours disulfide bonding between Cys32 and Cys60 (located at DI) and within Cys288 and Cys326 (located at DV) of β_2GP1. These bonds potentially encourage the binding of anti-β_2GP1 antibodies to β_2GP1, and might lead to thrombus formation. Oxidation and biotinylation of β_2GP1 glycan chains also induce β_2GP1 dimerisation, which raises β_2GP1 affinity [54]. Additionally, it is speculated that the intramolecular interaction and conformation of β_2GP1 can be affected by increased sialylation of β_2GP1 glycan structures [55].

Lastly, the structure of β_2GP1 can be inherently diverse. Among the four allelic variants, β_2GP1 Val/Val genotypes were frequently found to co-exist with anti-β_2GP1 antibodies [56]. It has also been proposed that the Val247 variant of circular-β_2GP1 is easier to transform into J-shaped β_2GP1 after losing the electrostatic interaction between Glu228 (located in DIV) and Lys308 (located in DV) [57]. Thus, this transformation exposes the cryptic epitope for antibody binding and raises the risk of thrombosis.

2.2. Physiological Role(s) of β_2GP1

The precise physiological role of β_2GP1 is unknown. β_2GP1-deficient individuals appear to be healthy, suggesting that β_2GP1 function might not be essential for life [58]. However, the disulphide bonds and phospholipid binding sites in β_2GP1 are highly conserved across the animal kingdom [36]. Therefore, it is very unlikely that this abundant and well-conserved molecule exists without a function.

Although β_2GP1-deficient individuals do not have an associated haemostatic abnormality, many functions in the regulation of haemostasis have been attributed to β_2GP1. First, β_2GP1 has been demonstrated to inhibit adenosine diphosphate (ADP)-mediated platelet aggregation and serotonin secretion [59,60]. Second, β_2GP1 might be a mediator for von Willebrand factor (vWF) activation and clearance. β_2GP1 has been reported to bind to the A1-domain of vWF, preferably vWF in a glycoprotein (GP) Ib-binding conformation. This low affinity binding allows the formation of disulfide bridges between β_2GP1 and vWF. Thus, the disulfide bridges prevent vWF-mediated platelet activation [15] and potentially protect the cleavage of vWF by the vWF protease, a disintegrin and metalloproteinase with a thrombospondin type 1 motif, member 13 (ADAMTS13) [61]. Thirdly, β_2GP1 has also been demonstrated to be involved in several coagulation pathways, yet these effects remain to be elucidated [60].

β_2GP1 has been suggested to be a general scavenger in circulation [62,63]. During apoptosis or cellular activation, the reorganisation of the plasma membrane exposes phosphatidylserine on the cell surface. β_2GP1 binds to phosphatidylserine expressed on these apoptotic cells [62], as well as platelet microparticles [63], to assist their phagocytosis by macrophages. In addition, β_2GP1 is also involved in innate immunity as demonstrated by the insertion of DV of β_2GP1 into bacterial membranes that can lead to cytosol leakage and death of bacteria [64]. β_2GP1 also changes its conformation while binding to lipopolysaccharide on Gram-negative bacteria, forming a complex which allows recognition and clearance by monocytes [65]. Finally, β_2GP1 might be important in embryonic development, as the percentage of null offspring born in β_2GP1 knock-out mice is lower than expected [66].

In summary, β_2GP1 has been proposed to be involved in a range of physiological processes, including clot formation, fibrinolysis, cell activation, immune responses, atherosclerosis, apoptosis, angiogenesis, and fetal loss [60]. Further research is clearly warranted to determine the precise physiological role(s) of β_2GP1.

3. Anti-β₂GP1 Antibodies

By itself, β₂GP1 has no deleterious effect on normal cellular function, but rather interferes with the physiological function of cells following binding with anti-β₂GP1 antibodies. Therefore, it has been proposed that anti-β₂GP1 antibodies induce a new function for β₂GP1 [67]. The affinity of β₂GP1 is low and only binds to anionic phospholipids below a certain concentration [41,48]. Upon binding with anionic phospholipids, it transforms into the J-shaped conformation and exposes the cryptic epitope located at DI which enables antibodies to bind. When the amount of β₂GP1 bound to anionic phospholipid membrane reaches a certain density, antibodies dimerise the adjacent β₂GP1 molecules [48]. This dimerisation forms a high affinity anti-β₂GP1-β₂GP1 complex, activating targeted cells and causing APL antibodies-related manifestations.

3.1. Clinical Significance of Anti-β₂GP1 Antibodies

The presence of anti-β₂GP1 antibodies, especially those with LAC activity, is highly associated with increased thrombotic risk compared to other APL antibody subgroups [10,11]. APS patients have higher levels of platelet activation as reflected by raised urinary thromboxane metabolites [68]. Moreover, the co-existence of J-shaped β₂GP1 and anti-β₂GP1 antibodies prolongs the activated partial thromboplastin time of normal plasma, compared to J-shaped β₂GP1 alone [41], suggesting that anti-β₂GP1 antibodies also affect secondary haemostasis. Conversely, 40% of APS patients have a prolonged bleeding time without an accompanying bleeding tendency [69]. Although there is no clear explanation for these contradictory findings, it suggests that anti-β₂GP1 antibodies affect normal haemostatic function.

The contribution of anti-β₂GP1 antibodies to placental-related pregnancy complications remains controversial. A systematic review and meta-analysis reported that there were insufficient data to support an association between anti-β₂GP1 antibodies and pregnancy complications [70]. However, an *in vitro* study demonstrated that anti-β₂GP1 antibodies stimulate trophoblasts to increase secretion of vascular endothelial growth factor, placental growth factor, and soluble endoglin, leading to a higher risk of obstetrical complication [71]. Furthermore, anti-β₂GP1-β₂GP1 complexes have been suggested to disrupt the anticoagulant shield formed by annexin A5 on vascular cells [72]. Thus, patients could be predisposed to placental thrombosis that may result in fetal growth restriction and/or pregnancy loss.

3.2. Etiology of Anti-β₂GP1 Antibodies

The etiology of anti-β₂GP1 antibodies remains unclear. Both genetic and environmental factors may contribute to their production [2,73]. Various animal models and family/population studies have indicated that several human leukocyte antigen genes are associated with the occurrence of APL antibodies and the development of thrombosis [74–76]. These pathogenic antibodies are thought to be produced by activated auto-reactive T and B cells due to the similarity between foreign and self-protein/peptide sequences (molecular mimicry) [77]. Viruses, bacteria, mycoplasma and parasites with the same amino acid sequences can also initiate antibody production [78]. However, this theory is unable to clearly explain the etiology, as antibodies are also produced by injecting anionic phospholipids such as cardiolipin, phosphatidylserine, or lipopolysaccharide into animals [79,80].

Anti-β₂GP1 antibodies might be naturally occurring antibodies, as benign and low affinity APL antibodies are found in 1%–5% of healthy individuals [3,81]. The mechanism(s) of transition of anti-β₂GP1 antibody from benign to pathogenic are unknown, however there is evidence to suggest that this may be induced by infection. β₂GP1 binds to pathogenic phospholipids such as protein H from *Streptococcus pyogenes* [82], causing conformational change, exposure of the cryptic epitope, and inducing production of pathogenic anti-DI-β₂GP1 antibodies. The conformation of β₂GP1 is also susceptible to many factors and may trigger the synthesis of antibodies. Similarly, antibody production can be prompted by ageing, vaccination, drugs, and malignancies. Their association with clinical manifestations, however, requires further investigation [2,73].

3.3. The Two Hit Hypothesis

The detection of anti-β_2GP1 antibodies in healthy individuals [3,4], APS, and SLE patients without complications [83] indicates that the antibody alone is insufficient for the pathogenesis of APS. It is proposed that a "first-hit" injury primes the endothelium, and a "second-hit" injury triggers thrombus formation. Studies have shown that anti-β_2GP1 antibodies infused into mice only initiate thrombus formation following vessel-wall injury [84,85]. Endothelium priming involves vessel-wall injury, infection, recent surgery [86], and rarely, the disturbance of redox balance in the vascular milieu [53]. Once primed, the "second-hit" injury, such as smoking, immobilisation, pregnancy, malignancy, *etc.*, stimulates the development of thrombosis [87].

3.4. Types of Anti-β_2GP1 Antibodies

The two hit hypothesis has been proposed to be a good model for the pathogenesis of APS [4]. Yet, it cannot clarify why APL antibodies present in healthy individuals are not pathogenic. Some studies suggest that this could be due to differences in the targeted epitope [10,48] and the structure of anti-β_2GP1 antibodies [4]. Anti-β_2GP1 antibodies isolated from primary APS patients are considered to be poly-reactive, as they have been found to react against several domains of β_2GP1, such as DV (52.9%–64.6%), DIV (45.8%), DI–II (33.1%), and DIII (20.5%) [88]. Anti-DI-β_2GP1 antibodies recognising the cryptic epitope of DI (Type A) in symptomatic APS patients are strongly associated with thrombotic history and positive LAC activity [10]. Conversely, antibodies that are directed against other domains (Type B) in healthy populations are weakly correlated with thrombosis. These more benign type B antibodies also have lower avidity compared to those pathogenic type A antibodies [89].

Besides binding epitopes, anti-β_2GP1 antibodies can be classified according to immunoglobulin (Ig) isotype; *i.e.*, IgG, IgM, and IgA. Among these, anti-β_2GP1 IgG antibodies are more strongly associated with the manifestations of APS [1]. Furthermore, different subclasses of anti-β_2GP1 IgG antibodies, predominantly IgG2 and IgG3, have also been identified in APS patients and healthy children, respectively [4]. IgG3 is the most effective activator for the classical complement pathway, hence leading to increased C3c (a complement component) activation and binding to anti-β_2GP1 IgG3 antibodies in healthy children [4]. Complement activation normally triggers platelet activation, which is related to the pathogenesis of APS [90,91]. However, C3c is an opsonin to improve the clearance of the bound target [92]. Instead of activating platelets, C3c binding enhances the clearance of pathogenic anti-β_2GP1 immune complexes and protects healthy children from complications. Moreover, anti-β_2GP1 antibodies in healthy and asymptomatic individuals are highly sialylated compared to symptomatic patients [4]. These sialylated anti-β_2GP1 antibodies have been found to have protective roles for healthy individuals because of their inability to bind and activate platelets.

3.5. Anti-DI-β_2GP1 Antibodies as a Diagnostic Tool

Anti-DI-β_2GP1 antibodies are highly associated with both vascular and obstetric complications, compared to antibodies against other domains of β_2GP1 [10]. Anti-DI-β_2GP1 antibodies are regularly isolated from APS patients compared to those with infection-induced transient APL antibody positivity. APS patients at higher risk of complications (triple APL positivity) also have higher titres of anti-DI-β_2GP1 antibodies [93], suggesting that the specificity of diagnosis of APS may increase when anti-DI-β_2GP1 antibodies are included. However, assays that detect anti-DI-β_2GP1 antibodies have lower sensitivity compared to those that detect the whole β_2GP1 molecule, as patients might produce clinically significant antibodies against other epitopes [46]. Currently, commercially available kits are not available for the detection of anti-DI-β_2GP1 antibodies. Instead, research assays with different sensitivities have been reported, such as ELISAs that use N-terminally biotinylated DI on streptavidin plates [94] and a β_2GP1-DI chemiluminescence immunoassay (CIA, INOVA Diagnostic, San Diego, CA, US) [95]. Further studies are warranted to determine the diagnostic and prognostic value of assays that detect anti-DI-β_2GP1 antibodies.

4. Anti-β_2GP1-β_2GP1 Complexes and Platelets

Although there is consensus that β_2GP1 interacts with anti-β_2GP1 antibodies to form anti-β_2GP1-β_2GP1 complexes with high affinity to anionic phospholipids [41,48], the affected pathway(s) remains unclear. Potential mechanisms by which APL antibodies might increase the risk of vascular and obstetric complications are reviewed elsewhere [13]. In this review, we have only focused on the effects of anti-β_2GP1 antibodies and β_2GP1 on platelets (Figure 2).

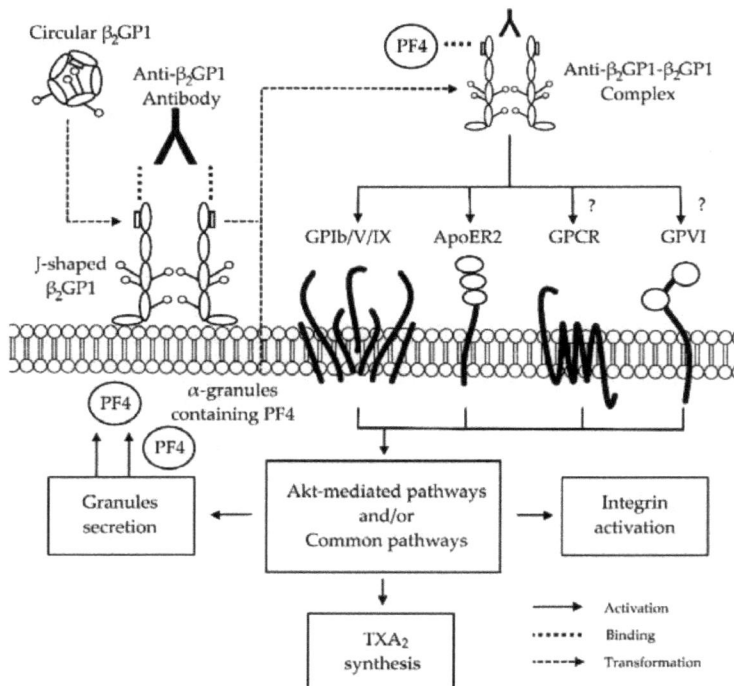

Figure 2. Proposed mechanisms of interaction between anti-beta-2-glycoprotein 1 (β_2GP1)-β_2GP1 complex and platelet receptors. Circular β_2GP1 binds to the anionic phospholipid platelet membrane and transforms into J-shaped β_2GP1. This allows the anti-domain 1-β_2GP1 antibody to bind and to form the anti-β_2GP1-β_2GP1 complex. The anti-β_2GP1-β_2GP1 complex has been proposed to interact with glycoprotein (GP) Ib of GPIb/V/IX [14] and apolipoprotein E receptor 2 (ApoER2) [96–98]. In our group, we propose that the complex might trigger adenosine diphosphate (ADP) and collagen-mediated pathways via guanine nucleotide-binding protein coupled receptor (GPCR) and GPVI, respectively [99,100]. Yet, further studies are needed to clarify the variability of results. The binding of the complex with receptors leads to the activation of protein kinase B (Akt)-mediated and/or common pathways, causing granules secretion, thromboxane A_2 (TXA$_2$) synthesis, integrin activation, and subsequently, clot formation. The platelet factor 4 (PF4) from secreted α-granules have also been showed to interact with the anti-β_2GP1-β_2GP1 complex [101] Abbreviations: β_2GP1, beta-2-glycoprotein 1; GP, glycoprotein; ApoER2, apolipoprotein E receptor 2; ADP, adenosine diphosphate; GPCR, guanine nucleotide-binding protein coupled receptor; TXA$_2$, thromboxane A_2; PF4, platelet factor 4; Akt, protein kinase B.

Platelets are a crucial component of haemostasis, a physiological process that forms a localised clot at the vessel injury site to limit blood loss while maintaining normal blood circulation [17,102]. Activation of platelet receptors leads to platelet adhesion, aggregation, activation of the protein kinase B-mediated and/or common pathways, secretion of granules, integrin activation, synthesis of thromboxane A_2, and finally, clot formation [17,103]. In the patients with autoimmune diseases, circular β_2GP1 transforms into the J-shaped conformation after binding to the phospholipid membrane of platelets, allowing anti-β_2GP1 antibodies to bind and form anti-β_2GP1-β_2GP1 complexes [48] (Figure 2). In turn, these complexes are proposed to activate platelet receptor(s)—e.g., glycoprotein (GP) Ib [14], apolipoprotein E receptor 2 (ApoER2) [16], guanine nucleotide-binding protein-coupled receptors-(GPCR) [100], and GPVI [99]. Furthermore, these complexes have also been suggested to affect other pathway(s) by inhibiting β_2GP1 binding to vWF [15] and by interacting with platelet factor 4 (PF4) secreted from platelets [101]. The activation of platelet receptor(s) by these mechanisms potentially results in excessive clot formation and/or pregnancy complications [14–16]. Therefore, understanding the effects of anti-β_2GP1-β_2GP1 complexes on platelets is important not only to determine the mechanism(s) of interaction, but to also potentially assist in the development of novel or improved treatments for patients with autoimmune diseases.

It has been reported that β_2GP1 directly binds to GPIb of the GPIb/V/IX receptor via DII–V [14]. The presence of anti-DI-β_2GP1 antibodies potentially dimerises β_2GP1 and inappropriately initiates GPIb-mediated platelet adhesion and aggregation [14,104]. This activation by anti-β_2GP1-β_2GP1 complexes may explain the increased thrombotic risk in APS patients [14].

Besides the GPIb receptor, DV of β_2GP1 has been shown to dimerise and interact with the A1 portion of ApoER2 [96–98]. ApoER2, also known as low-density lipoprotein receptor-related protein 8, is the only low-density lipoprotein family receptor found on platelets [96]. This receptor is recognised to be targeted by the anti-β_2GP1-β_2GP1 complex, as the blockage of ApoER2 by its antagonist diminishes the effect of the anti-β_2GP1-β_2GP1 complex to increase the adhesion of platelets to collagen [105]. It has also been established that the interaction of anti-β_2GP1-β_2GP1 complexes with ApoER2 activates platelet analogously to GPIb-mediated platelet activation [16]. Recently, a dimer composed of two A1 portions of ApoER2 joined by a flexible link has been created [98]. This dimer is able to inhibit anti-β_2GP1-β_2GP1 complexes from binding to negatively-charged phospholipids and ApoER2 [98], reflecting another possible treatment option for patients with APS.

Anti-β_2GP1-β_2GP1 complexes may also affect GPCR and GPVI-mediated platelet activation pathways. Anti-β_2GP1 antibodies from different origins have recently been reported to exhibit diverse effects on *in vitro* platelet aggregation. Affinity purified rabbit [99] and SLE patient-derived anti-β_2GP1 antibodies [100] demonstrated inhibitory and enhancement effects, respectively, on ADP-induced platelet aggregation. When collagen was used, affinity purified rabbit anti-β_2GP1 antibodies [99] enhanced platelet aggregation. However, no effect was demonstrated using patient-derived IgG fractions (containing aCL and anti-β_2GP1 antibodies) [106] and affinity-purified goat anti-β_2GP1 antibodies [107]. Based on these results, it is difficult to arrive at a consensus due to the variable effects possibly caused by anti-β_2GP1 antibodies with different structure and binding specificities. Thus, further research is needed to elucidate the variable effects of anti-β_2GP1-β_2GP1 complexes on GPCR- and GPVI-mediated pathways.

As described above, β_2GP1 binds with vWF to prevent platelet activation. It has been suggested that anti-β_2GP1 antibodies in APS patients can neutralise this inhibitory effect, potentially leading to thrombosis and consumptive thrombocytopenia [15]. Furthermore, PF4, a pro-coagulant factor secreted from the α granules of platelets, has also been demonstrated to interact with β_2GP1 [101]. PF4 is proposed to dimerise and stabilise β_2GP1 on phospholipids, ensuring that β_2GP1 is easily recognised by anti-β_2GP1 antibodies. The formation of anti-β_2GP1-β_2GP1-PF4 complexes may activate platelets, leading to the development of thrombosis in APS patients [101].

5. Conclusion and Further Research

There is substantial literature available on the interaction between three interchangeable β_2GP1 structures and anti-β_2GP1 antibodies. The transformation of S-shaped or circular β_2GP1 to J-shaped β_2GP1 exposes the cryptic epitope in DI, enabling the binding of anti-β_2GP1 antibodies, particularly those to DI of β_2GP1. The formation of the anti-β_2GP1-β_2GP1 complex is thought to be responsible for the increased risk of thrombosis and/or pregnancy complications in patients with autoimmune diseases. Although numerous mechanisms of interaction between anti-β_2GP1-β_2GP1 complex and receptors/components have been proposed, the actual affected physiological pathway(s) remain unclear. One of the possible explanations for these ambiguities is the use of anti-β_2GP1 antibodies with different structures and binding specificities from patient- and animal-derived origins across different studies. Therefore, further research is required to better clarify and categorise the type of antibodies used. This approach will in turn facilitate studies that will lead to increased understanding of the interactions between these antibodies and platelets.

In conclusion, the standardisation and development of methods, such as anti-DI-β_2GP1 antibody ELISAs, are required to differentiate between the types and pathogenicity of anti-β_2GP1 antibodies. This will allow more meaningful interpretation of laboratory- and clinic-based findings, which will potentially lead to the elucidation of the mechanism(s) of interaction between β_2GP1, anti-β_2GP1 antibodies and platelets. In combination, these further developments can help to improve the diagnostic and therapeutic techniques for patients with APS, and perhaps more widely, autoimmune diseases.

Acknowledgments: The authors acknowledge the support of the Lupus Association of Tasmania and the Clifford Craig Research Trust.

Author Contributions: Yik C. Ho, Kiran D. K. Ahuja, Heinrich Körner and Murray J. Adams wrote the review.

Conflicts of Interest: The authors declare no conflict of interest.

Abbreviations

The following abbreviations are used in this manuscript:

APL	Anti-phospholipid
LAC	Lupus anti-coagulant
aCL	Anti-cardiolipin
Anti-β_2GP1	Anti-beta 2 glycoprotein 1
APS	Antiphospholipid syndrome
SLE	Systemic lupus erythematosus
ELISA	Enzyme-linked immunosorbent assays
D	Domain
Anti-DI-β_2GP1	Anti-domain I-beta 2 glycoprotein 1
EDTA	Ethylenediaminetetraacetic acid
ADP	Adenosine diphosphate
VWF	Von Willebrand factor
GP	Glycoprotein
ADAMTS13	vWF protease
Ig	Immunoglobulin
ApoER2	Apolipoprotein E receptor 2
GPCR	Guanine nucleotide-binding protein-coupled receptors
PF4	Platelet factor 4

References

1. Keeling, D.; Mackie, I.; Moore, G.W.; Greer, I.A.; Greaves, M. Guidelines on the investigation and management of antiphospholipid syndrome. *Br. J. Haematol.* **2012**, *157*, 47–58. [CrossRef] [PubMed]
2. Biggioggero, M.; Meroni, P.L. The geoepidemiology of the antiphospholipid antibody syndrome. *Autoimmun. Rev.* **2010**, *9*, A299–A304. [CrossRef] [PubMed]
3. De Groot, P.G.; Urbanus, R.T. The significance of autoantibodies against β_2-glycoprotein I. *Blood* **2012**, *120*, 266–274. [CrossRef] [PubMed]

4. Fickentscher, C.; Magorivska, I.; Janko, C.; Biermann, M.; Bilyy, R.; Nalli, C.; Tincani, A.; Medeghini, V.; Meini, A.; Nimmerjahn, F. The pathogenicity of anti-β$_2$GP1-IgG autoantibodies depends on Fc glycosylation. *J. Immunol. Res.* **2015**, *2015*, 1–15. [CrossRef]

5. Miyakis, S.; Lockshin, M.; Atsumi, T.; Branch, D.; Brey, R.; Cervera, R.; Derksen, R.; De Groot, P.; Koike, T.; Meroni, P. International consensus statement on an update of the classification criteria for definite antiphospholipid syndrome (APS). *J. Thromb. Haemost.* **2006**, *4*, 295–306. [CrossRef] [PubMed]

6. Urbanus, R.T.; Derksen, R.H.; De Groot, P.G. Current insight into diagnostics and pathophysiology of the antiphospolipid syndrome. *Blood Rev.* **2008**, *22*, 93–105. [CrossRef] [PubMed]

7. De Groot, P.G.; Urbanus, R.T. The future of antiphospholipid antibody testing. *Semin. Thromb. Hemost.* **2012**, *38*, 412–420. [CrossRef] [PubMed]

8. Reber, G.; Boehlen, F.; De Moerloose, P. Technical aspects in laboratory testing for antiphospholipid antibodies: Is standardization an impossible dream? *Semin. Thromb. Hemost.* **2008**, *34*, 340–346. [CrossRef] [PubMed]

9. Du, V.X.; Kelchtermans, H.; De Groot, P.G.; De Laat, B. From antibody to clinical phenotype, the black box of the antiphospholipid syndrome: Pathogenic mechanisms of the antiphospholipid syndrome. *Thromb. Res.* **2013**, *132*, 319–326. [CrossRef] [PubMed]

10. De Laat, B.; Derksen, R.H.; Urbanus, R.T.; De Groot, P.G. IgG antibodies that recognize epitope Gly40-Arg43 in domain I of β$_2$–glycoprotein I cause LAC, and their presence correlates strongly with thrombosis. *Blood* **2005**, *105*, 1540–1545. [CrossRef] [PubMed]

11. De Laat, B.; Derksen, R.; Reber, G.; Musial, J.; Swadzba, J.; Bozic, B.; Cucnik, S.; Regnault, V.; Forastiero, R.; Woodhams, B. An international multicentre-laboratory evaluation of a new assay to detect specifically lupus anticoagulants dependent on the presence of anti-beta2-glycoprotein I autoantibodies. *J. Thromb. Haemost.* **2011**, *9*, 149–153. [CrossRef] [PubMed]

12. Chamley, L.; McKay, E.; Pattison, N. Cofactor dependent and cofactor independent anticardiolipin antibodies. *Thromb. Res.* **1991**, *61*, 291–299. [CrossRef]

13. Tripodi, A.; De Groot, P.G.; Pengo, V. Antiphospholipid syndrome: Laboratory detection, mechanisms of action and treatment. *J. Intern. Med.* **2011**, *270*, 110–122. [CrossRef] [PubMed]

14. Shi, T.; Giannakopoulos, B.; Yan, X.; Yu, P.; Berndt, M.C.; Andrews, R.K.; Rivera, J.; Iverson, G.M.; Cockerill, K.A.; Linnik, M.D.; *et al.* Anti-β$_2$-glycoprotein I antibodies in complex with β$_2$-glycoprotein I can activate platelets in a dysregulated manner via glycoprotein Ib-IX-V. *Arthritis Rheum.* **2006**, *54*, 2558–2567. [CrossRef] [PubMed]

15. Hulstein, J.J.; Lenting, P.J.; De Laat, B.; Derksen, R.H.; Fijnheer, R.; De Groot, P.G. β$_2$-glycoprotein I inhibits von Willebrand factor–dependent platelet adhesion and aggregation. *Blood* **2007**, *110*, 1483–1491. [CrossRef] [PubMed]

16. Korporaal, S.J.; Relou, I.A.; Van Eck, M.; Strasser, V.; Bezemer, M.; Gorter, G.; Van Berkel, T.J.; Nimpf, J.; Akkerman, J.-W.N.; Lenting, P.J. Binding of low density lipoprotein to platelet apolipoprotein E receptor 2' results in phosphorylation of p38MAPK. *J. Biol. Chem.* **2004**, *279*, 52526–52534. [CrossRef] [PubMed]

17. Jurk, K.; Kehrel, B.E. Platelets: Physiology and biochemistry. *Semin. Thromb. Hemost.* **2005**, *31*, 381–392. [CrossRef] [PubMed]

18. Galli, M.; Luciani, D.; Bertolini, G.; Barbui, T. Lupus anticoagulants are stronger risk factors for thrombosis than anticardiolipin antibodies in the antiphospholipid syndrome: A systematic review of the literature. *Blood* **2003**, *101*, 1827–1832. [CrossRef] [PubMed]

19. Galli, M.; Barbui, T. Antiphospholipid antibodies and pregnancy. *Best Pract. Res. Clin. Haematol.* **2003**, *16*, 211–225. [CrossRef]

20. Favaloro, E.J.; Silvestrini, R. Assessing the usefulness of anticardiolipin antibody assays. *Am. J. Clin. Pathol.* **2002**, *118*, 548–557. [CrossRef] [PubMed]

21. Bahar, A.; Kwak, J.; Beer, A.; Kim, J.; Nelson, L.; Beaman, K.; Gilman-Sachs, A. Antibodies to phospholipids and nuclear antigens in non-pregnant women with unexplained spontaneous recurrent abortions. *J. Reprod. Immunol.* **1993**, *24*, 213–222. [CrossRef]

22. Spadaro, A.; Riccieri, V.; Terracina, S.; Rinaldi, T.; Taccari, E.; Zoppinia, A. Class specific rheumatoid factors and antiphospholipid syndrome in systemic lupus erythematosus. *Lupus* **2000**, *9*, 56–60. [CrossRef]

23. Reber, G.; de Moerloose, P. Anti-β$_2$-glycoprotein I antibodies—When and how should they be measured? *Thromb. Res.* **2004**, *114*, 527–531. [CrossRef]

24. Faden, D.; Tincani, A.; Tanzi, P.; Spatola, L.; Lojacono, A.; Tarantini, M.; Balestrieri, G. Anti-beta 2 glycoprotein I antibodies in a general obstetric population: Preliminary results on the prevalence and correlation with pregnancy outcome. Anti-beta2 glycoprotein I antibodies are associated with some obstetrical complications, mainly preeclampsia-eclampsia. *Eur. J. Obstet. Gynecol. Reprod. Biol.* **1997**, *73*, 37–42. [PubMed]

25. Atsumi, T.; Ieko, M.; Bertolaccini, M.L.; Ichikawa, K.; Tsutsumi, A.; Matsuura, E.; Koike, T. Association of autoantibodies against the phosphatidylserine-prothrombin complex with manifestations of the antiphospholipid syndrome and with the presence of lupus anticoagulant. *Arthritis Rheum.* **2000**, *43*, 1982–1993. [CrossRef]

26. McIntyre, J.A.; Wagenknecht, D.R.; Sugi, T. Phospholipid binding plasma proteins required for antiphospholipid antibody detection—An overview. *Am. J. Reprod. Immunol.* **1997**, *37*, 101–110. [CrossRef]

27. Matsuura, E.; Igarashi, Y.; Fujimoto, M.; Ichikawa, K.; Koike, T. Anticardiolipin cofactor(s) and differential diagnosis of autoimmune disease. *Lancet* **1990**, *336*, 177–178. [CrossRef]

28. McNeil, H.P.; Simpson, R.J.; Chesterman, C.N.; Krilis, S.A. Anti-phospholipid antibodies are directed against a complex antigen that includes a lipid-binding inhibitor of coagulation: Beta 2-glycoprotein I (apolipoprotein H). *Proc. Natl. Acad. Sci. USA.* **1990**, *87*, 4120–4124. [CrossRef] [PubMed]

29. Galli, M.; Comfurius, P.; Maassen, C.; Hemker, H.C.; de Baets, M.H.; van Breda-Vriesman, P.J.; Barbui, T.; Zwaal, R.F.; Bevers, E.M. Anticardiolipin antibodies (ACA) directed not to cardiolipin but to a plasma protein cofactor. *Lancet* **1990**, *335*, 1544–1547. [CrossRef]

30. Schultze, H. Glycoproteins of human plasma. *Bull. Schweiz. Akad. Med. Wiss.* **1961**, *17*, 77–91. [PubMed]

31. Lozier, J.; Takahashi, N.; Putnam, F.W. Complete amino acid sequence of human plasma beta 2-glycoprotein I. *Proc. Natl. Acad. Sci. USA.* **1984**, *81*, 3640–3644. [CrossRef] [PubMed]

32. Polz, E.; Kostner, G.M. The binding of β_2-glycoprotein I to human serum lipoproteins: Distribution among density fractions. *FEBS Lett.* **1979**, *102*, 183–186. [CrossRef]

33. Vlachoyiannopoulos, P.; Krilis, S.; Hunt, J.; Manoussakis, M.; Moutsopoulos, H. Patients with anticardiolipin antibodies with and without antiphospholipid syndrome: Their clinical features and β_2-glycoprotein I plasma levels. *Eur. J. Clin. Invest.* **1992**, *22*, 482–487. [CrossRef] [PubMed]

34. Miyakis, S.; Giannakopoulos, B.; Krilis, S.A. Beta 2 glycoprotein I-function in health and disease. *Thromb. Res.* **2004**, *114*, 335–346. [CrossRef] [PubMed]

35. Ioannou, Y.; Rahman, A. Domain I of β_2-glycoprotein I: Its role as an epitope and the potential to be developed as a specific target for the treatment of the antiphospholipid syndrome. *Lupus* **2010**, *19*, 400–405. [CrossRef]

36. Agar, C.; De Groot, P.G.; Marquart, J.A.; Meijers, J. Evolutionary conservation of the lipopolysaccharide binding site of β_2-glycoprotein I. *Thromb. Haemost.* **2011**, *106*, 1069–1075. [CrossRef] [PubMed]

37. Steinkasserer, A.; Estaller, C.; Weiss, E.H.; Sim, R.B.; Day, A.J. Complete nucleotide and deduced amino acid sequence of human beta 2-glycoprotein I. *Biochem. J.* **1991**, *277*, 387–391. [CrossRef] [PubMed]

38. Bouma, B.; De Groot, P.G.; Van Den Elsen, J.M.; Ravelli, R.B.; Schouten, A.; Simmelink, M.J.; Derksen, R.H.; Kroon, J.; Gros, P. Adhesion mechanism of human β_2-glycoprotein I to phospholipids based on its crystal structure. *EMBO J.* **1999**, *18*, 5166–5174. [CrossRef]

39. Pelkmans, L.; De Laat, B. Antibodies against domain I of β_2-glycoprotein I: The one and only? *Lupus* **2012**, *21*, 769–772. [CrossRef] [PubMed]

40. Kristensen, T.; Schousboe, I.; Boel, E.; Mulvihill, E.M.; Hansen, R.R.; Moller, K.B.; Moller, N.P.H.; Sottrup-Jensen, L. Molecular cloning and mammalian expression of human β_2-glycoprotein I cDNA. *FEBS Lett.* **1991**, *289*, 183–186. [CrossRef]

41. Agar, C.; Van Os, G.M.; Morgelin, M.; Sprenger, R.R.; Marquart, J.A.; Urbanus, R.T.; Derksen, R.H.; Meijers, J.C.; De Groot, P.G. β_2-Glycoprotein I can exist in two conformations: Implications for our understanding of the antiphospholipid syndrome. *Blood* **2010**, *116*, 1336–1343. [CrossRef] [PubMed]

42. Schwarzenbacher, R.; Zeth, K.; Diederichs, K.; Gries, A.; Kostner, G.M.; Laggner, P.; Prassl, R. Crystal structure of human β_2-glycoprotein I: Implications for phospholipid binding and the antiphospholipid syndrome. *EMBO J.* **1999**, *18*, 6228–6239. [CrossRef] [PubMed]

43. Hammel, M.; Kriechbaum, M.; Gries, A.; Kostner, G.M.; Laggner, P.; Prassl, R. Solution structure of human and bovine β_2-glycoprotein I revealed by small-angle X-ray scattering. *J. Mol. Biol.* **2002**, *321*, 85–97. [CrossRef]

44. Koike, T.; Ichikawa, K.; Kasahara, H.; Atsumi, T.; Tsutsumi, A.; Matsuura, E. Epitopes on β_2GPI recognized by anticardiolipin antibodies. *Lupus* **1998**, *7*, 14–17. [CrossRef]

45. Ioannou, Y.; Pericleous, C.; Giles, I.; Latchman, D.S.; Isenberg, D.A.; Rahman, A. Binding of antiphospholipid antibodies to discontinuous epitopes on domain I of human β_2-glycoprotein I: Mutation studies including residues R39 to R43. *Arthritis Rheum.* **2007**, *56*, 280–290. [CrossRef] [PubMed]

46. Chighizola, C.B.; Gerosa, M.; Meroni, P.L. New tests to detect antiphospholipid antibodies: Anti-domain I beta-2-glycoprotein-I antibodies. *Curr. Rheumatol. Rep.* **2014**, *16*, 1–9. [CrossRef] [PubMed]

47. De Laat, B.; De Groot, P.G. Autoantibodies directed against domain I of beta2-glycoprotein I. *Curr. Rheumatol. Rep.* **2011**, *13*, 70–76. [CrossRef] [PubMed]

48. De Laat, B.; Derksen, R.H.W.M.; Van Lummel, M.; Pennings, M.T.T.; De Groot, P.G. Pathogenic anti-β_2-glycoprotein I antibodies recognize domain I of β_2-glycoprotein I only after a conformational change. *Blood* **2006**, *107*, 1916–1924. [CrossRef] [PubMed]

49. Wurm, H. β_2-Glycoprotein-I (apolipoprotein H) interactions with phospholipid vesicles. *Int. J. Biochem.* **1984**, *16*, 511–515. [CrossRef]

50. Perutková, Š.; Frank-Bertoncelj, M.; Rozman, B.; Kralj-Iglič, V.; Iglič, A. Influence of ionic strength and beta2-glycoprotein I concentration on agglutination of like-charged phospholipid membranes. *Coll. Surf. B* **2013**, *111*, 699–706. [CrossRef] [PubMed]

51. Hoshino, M.; Hagihara, Y.; Nishii, I.; Yamazaki, T.; Kato, H.; Goto, Y. Identification of the phospholipid-binding site of human β_2-glycoprotein I domain V by heteronuclear magnetic resonance. *J. Mol. Biol.* **2000**, *304*, 927–939. [CrossRef] [PubMed]

52. Jankowski, M.; Vreys, I.; Wittevrongel, C.; Boon, D.; Vermylen, J.; Hoylaerts, M.F.; Arnout, J. Thrombogenicity of β_2-glycoprotein I–dependent antiphospholipid antibodies in a photochemically induced thrombosis model in the hamster. *Blood* **2003**, *101*, 157–162. [CrossRef] [PubMed]

53. Giannakopoulos, B.; Krilis, S.A. The pathogenesis of the antiphospholipid syndrome. *N. Engl. J. Med.* **2013**, *368*, 1033–1044. [CrossRef] [PubMed]

54. Dupuy d'Angeac, A.; Stefas, I.; Graafland, H.; de Lamotte, F.; Rucheton, M.; Palais, C.; Eriksson, A.-K.; Bosc, P.; Rosé, C.; Chicheportiche, R. Biotinylation of glycan chains in β_2 glycoprotein I induces dimerization of the molecule and its detection by the human autoimmune anti-cardiolipin antibody EY2C9. *Biochem. J.* **2006**, *393*, 117–127. [CrossRef]

55. Kondo, A.; Miyamoto, T.; Yonekawa, O.; Giessing, A.M.; Østerlund, E.C.; Jensen, O.N. Glycopeptide profiling of beta-2-glycoprotein I by mass spectrometry reveals attenuated sialylation in patients with antiphospholipid syndrome. *J. Proteomics* **2009**, *73*, 123–133. [CrossRef] [PubMed]

56. Chamorro, A.-J.; Marcos, M.; Mirón-Canelo, J.-A.; Cervera, R.; Espinosa, G. Val247Leu beta2-glycoprotein-I allelic variant is associated with antiphospholipid syndrome: Systematic review and meta-analysis. *Autoimmun. Rev.* **2012**, *11*, 705–712. [CrossRef] [PubMed]

57. Yasuda, S.; Atsumi, T.; Matsuura, E.; Kaihara, K.; Yamamoto, D.; Ichikawa, K.; Koike, T. Significance of valine/leucine[247] polymorphism of β_2-glycoprotein I in antiphospholipid syndrome: Increased reactivity of anti-β_2-glycoprotein I autoantibodies to the valine[247] β_2-glycoprotein I variant. *Arthritis Rheum.* **2005**, *52*, 212–218. [CrossRef] [PubMed]

58. Haupt, H.; Schwick, H.G.; Storiko, K. On a hereditary beta-2-glycoprotein I deficiency. *Humangenetik* **1968**, *5*, 291–293. [PubMed]

59. Nimpf, J.; Wurm, H.; Kostner, G. Interaction of beta 2-glycoprotein-I with human blood platelets: Influence upon the ADP-induced aggregation. *Thromb. Haemost.* **1985**, *54*, 397–401. [PubMed]

60. Rahgozar, S. Revisiting Beta 2 glycoprotein I, the major autoantigen in the antiphospholipid syndrome. *Iran. J. Immunol.* **2012**, *9*, 73–85. [PubMed]

61. Passam, F.; Rahgozar, S.; Qi, M.; Raftery, M.; Wong, J.; Tanaka, K.; Ioannou, Y.; Zhang, J.; Gemmell, R.; Qi, J. Redox control of β_2-glycoprotein I–von Willebrand factor interaction by thioredoxin-1. *J. Thromb. Haemost.* **2010**, *8*, 1754–1762. [CrossRef] [PubMed]

62. Maiti, S.N.; Balasubramanian, K.; Ramoth, J.A.; Schroit, A.J. Beta-2-glycoprotein 1-dependent macrophage uptake of apoptotic cells. Binding to lipoprotein receptor-related protein receptor family members. *J. Biol. Chem.* **2008**, *283*, 3761–3766. [CrossRef] [PubMed]

63. Abdel-Monem, H.; Dasgupta, S.K.; Le, A.; Prakasam, A.; Thiagarajan, P. Phagocytosis of platelet microvesicles and β_2-glycoprotein I. *Thromb. Haemost.* **2010**, *104*, 335–341. [CrossRef] [PubMed]

64. Nilsson, M.; Wasylik, S.; Morgelin, M.; Olin, A.I.; Meijers, J.; Derksen, R.H.; De Groot, P.G.; Herwald, H. The antibacterial activity of peptides derived from human β_2 glycoprotein I is inhibited by protein H and M1 protein from Streptococcus pyogenes. *Mol. Microbiol.* **2008**, *67*, 482–492. [CrossRef] [PubMed]

65. Agar, C.; De Groot, P.G.; Morgelin, M.; Monk, S.D.; Van Os, G.; Levels, J.H.; De Laat, B.; Urbanus, R.T.; Herwald, H.; Van der Poll, T.; *et al.* β_2-glycoprotein I: A novel component of innate immunity. *Blood* **2011**, *117*, 6939–6947. [CrossRef] [PubMed]

66. Sheng, Y.; Reddel, S.W.; Herzog, H.; Wang, Y.X.; Brighton, T.; France, M.P.; Robertson, S.A.; Krilis, S.A. Impaired thrombin generation in β_2-glycoprotein I null mice. *J. Biol. Chem.* **2001**, *276*, 13817–13821. [PubMed]

67. De Groot, P.G.; Derksen, R. Pathophysiology of antiphospholipid antibodies. *Neth. J. Med.* **2004**, *62*, 267–272. [PubMed]

68. Forastiero, R.; Martinuzzo, M.; Carreras, L.O.; Maclouf, J. Anti-β_2 glycoprotein I antibodies and platelet activation in patients with antiphospholipid antibodies: Association with increased excretion of platelet-derived thromboxane urinary metabolites. *Thromb. Haemost.* **1998**, *79*, 42–45. [PubMed]

69. Urbanus, R.T.; De Laat, H.B.; De Groot, P.G.; Derksen, R.H. Prolonged bleeding time and lupus anticoagulant: A second paradox in the antiphospholipid syndrome. *Arthritis Rheum.* **2004**, *50*, 3605–3609. [CrossRef] [PubMed]

70. Abou-Nassar, K.; Carrier, M.; Ramsay, T.; Rodger, M.A. The association between antiphospholipid antibodies and placenta mediated complications: A systematic review and meta-analysis. *Thromb. Res.* **2011**, *128*, 77–85. [CrossRef] [PubMed]

71. Carroll, T.Y.; Mulla, M.J.; Han, C.S.; Brosens, J.J.; Chamley, L.W.; Giles, I.; Pericleous, C.; Rahman, A.; Sfakianaki, A.K.; Paidas, M.J. Modulation of trophoblast angiogenic factor secretion by antiphospholipid antibodies is not reversed by heparin. *Am. J. Reprod. Immunol.* **2011**, *66*, 286–296. [CrossRef] [PubMed]

72. Rand, J.; Wu, X.; Quinn, A.; Taatjes, D. The annexin A5-mediated pathogenic mechanism in the antiphospholipid syndrome: Role in pregnancy losses and thrombosis. *Lupus* **2010**, *19*, 460–469. [CrossRef] [PubMed]

73. Willis, R.; Shoenfeld, Y.; Pierangeli, S.S.; Blank, M. *What is the origin of antiphospholipid antibodies? In Antiphospholipid Syndrome*; Springer: New York, NY, USA, 2012; pp. 23–39.

74. Hashimoto, Y.; Kawamura, M.; Ichikawa, K.; Suzuki, T.; Sumida, T.; Yoshida, S.; Matsuura, E.; Ikehara, S.; Koike, T. Anticardiolipin antibodies in NZW × BXSB F1 mice. A model of antiphospholipid syndrome. *J. Immunol.* **1992**, *149*, 1063–1068. [PubMed]

75. Ida, A.; Hirose, S.; Hamano, Y.; Kodera, S.; Jiang, Y.; Abe, M.; Zhang, D.; Nishimura, H.; Shirai, T. Multigenic control of lupus-associated antiphospholipid syndrome in a model of (NZW × BXSB) F1 mice. *Eur. J. Immunol.* **1998**, *28*, 2694–2703. [CrossRef]

76. Castro-Marrero, J.; Balada, E.; Vilardell-Tarres, M.; Ordi-Ros, J. Genetic risk factors of thrombosis in the antiphospholipid syndrome. *Br. J. Haematol.* **2009**, *147*, 289–296. [CrossRef] [PubMed]

77. Epstein, F.H.; Albert, L.J.; Inman, R.D. Molecular mimicry and autoimmunity. *N. Engl. J. Med.* **1999**, *341*, 2068–2074. [CrossRef] [PubMed]

78. Sherer, Y.; Blank, M.; Shoenfeld, Y. Antiphospholipid syndrome (APS): Where does it come from? *Best Pract. Res. Clin. Rheumatol.* **2007**, *21*, 1071–1078. [CrossRef] [PubMed]

79. Gotoh, M.; Matsuda, J. Induction of anticardiolipin antibody and/or lupus anticoagulant in rabbits by immunization with lipoteichoic acid, lipopolysaccharide and lipid A. *Lupus* **1996**, *5*, 593–597. [CrossRef] [PubMed]

80. Subang, R.; Levine, J.S.; Janoff, A.S.; Davidson, S.M.; Taraschi, T.F.; Koike, T.; Minchey, S.R.; Whiteside, M.; Tannenbaum, M.; Rauch, J. Phospholipid-bound β_2-glycoprotein I induces the production of anti-phospholipid antibodies. *J. Autoimmun.* **2000**, *15*, 21–32. [CrossRef] [PubMed]

81. Merrill, J.T. Do antiphospholipid antibodies develop for a purpose? *Curr. Rheumatol. Rep.* **2006**, *8*, 109–113. [CrossRef] [PubMed]

82. Van Os, G.M.; Meijers, J.C.; Agar, C.; Seron, M.V.; Marquart, J.A.; Akesson, P.; Urbanus, R.T.; Derksen, R.H.; Herwald, H.; Morgelin, M.; *et al.* Induction of anti-β_2-glycoprotein I autoantibodies in mice by protein H of Streptococcus pyogenes. *J. Thromb. Haemost.* **2011**, *9*, 2447–2456. [CrossRef] [PubMed]

83. Biasiolo, A.; Rampazzo, P.; Brocco, T.; Barbero, F.; Rosato, A.; Pengo, V. [Anti-β2 Glycoprotein I—β2 Glycoprotein I] immune complexes in patients with antiphospholipid syndrome and other autoimmune diseases. *Lupus* **1999**, *8*, 121–126. [CrossRef] [PubMed]

84. Fischetti, F.; Durigutto, P.; Pellis, V.; Debeus, A.; Macor, P.; Bulla, R.; Bossi, F.; Ziller, F.; Sblattero, D.; Meroni, P. Thrombus formation induced by antibodies to β2-glycoprotein I is complement dependent and requires a priming factor. *Blood* **2005**, *106*, 2340–2346. [CrossRef] [PubMed]

85. Arad, A.; Proulle, V.; Furie, R.A.; Furie, B.C.; Furie, B. β2-glycoprotein-1 autoantibodies from patients with antiphospholipid syndrome are sufficient to potentiate arterial thrombus formation in a mouse model. *Blood* **2011**, *117*, 3453–3459. [CrossRef] [PubMed]

86. Asherson, R.A. The catastrophic antiphospholipid syndrome, 1998. A review of the clinical features, possible pathogenesis and treatment. *Lupus* **1998**, *7*, S55–S62. [CrossRef] [PubMed]

87. Agarwal, M.B. Antiphospholipid syndrome. *East. J. Med.* **2009**, *14*, 51–56.

88. Shoenfeld, Y.; Krause, I.; Kvapil, F.; Sulkes, J.; Lev, S.; Von Landenberg, P.; Font, J.; Zaech, J.; Cervera, R.; Piette, J. Prevalence and clinical correlations of antibodies against six β2-glycoprotein-I-related peptides in the antiphospholipid syndrome. *J. Clin. Immunol.* **2003**, *23*, 377–383. [CrossRef] [PubMed]

89. Cucnik, S.; Kveder, T.; Artenjak, A.; Gallova, Z.U.; Swadzba, J.; Musial, J.; Iwaniec, T.; Stojanovich, L.; Alessandri, C.; Valesini, G. Avidity of anti-β2-glycoprotein I antibodies in patients with antiphospholipid syndrome. *Lupus* **2012**, *21*, 764–765. [CrossRef] [PubMed]

90. Carter, A.M. Complement activation: An emerging player in the pathogenesis of cardiovascular disease. *Scientifica* **2012**, *2012*, 1–14. [CrossRef] [PubMed]

91. Jefferis, R.; Kumararatne, D. Selective IgG subclass deficiency: Quantification and clinical relevance. *Clin. Exp. Immunol.* **1990**, *81*, 357. [CrossRef] [PubMed]

92. Palarasah, Y.; Skjodt, K.; Brandt, J.; Teisner, B.; Koch, C.; Vitved, L.; Skjoedt, M.O. Generation of a C3c specific monoclonal antibody and assessment of C3c as a putative inflammatory marker derived from complement factor C3. *J. Immunol. Methods* **2010**, *362*, 142–150. [CrossRef] [PubMed]

93. Banzato, A.; Pozzi, N.; Frasson, R.; De Filippis, V.; Ruffatti, A.; Bison, E.; Padayattil, S.; Denas, G.; Pengo, V. Antibodies to domain I of β2 glycoprotein I are in close relation to patients risk categories in antiphospholipid syndrome (APS). *Thromb. Res.* **2011**, *128*, 583–586. [CrossRef] [PubMed]

94. Pozzi, N.; Banzato, A.; Bettin, S.; Bison, E.; Pengo, V.; De Filippis, V. Chemical synthesis and characterization of wild-type and biotinylated N-terminal domain 1–64 of beta2-glycoprotein I. *Protein Sci.* **2010**, *19*, 1065–1078. [CrossRef] [PubMed]

95. Meneghel, L.; Ruffatti, A.; Gavasso, S.; Tonello, M.; Mattia, E.; Spiezia, L.; Tormene, D.; Hoxha, A.; Fedrigo, M.; Simioni, P. Detection of IgG anti-domain I beta2 glycoprotein I antibodies by chemiluminescence immunoassay in primary antiphospholipid syndrome. *Clin. Chim. Acta* **2015**, *446*, 201–205. [CrossRef] [PubMed]

96. Van Lummel, M.; Pennings, M.T.; Derksen, R.H.; Urbanus, R.T.; Lutters, B.C.; Kaldenhoven, N.; De Groot, P.G. The binding site in β2-glycoprotein I for ApoER2' on platelets is located in domain V. *J. Biol. Chem.* **2005**, *280*, 36729–36736. [CrossRef] [PubMed]

97. Pennings, M.T.; Derksen, R.H.; Urbanus, R.T.; Tekelenburg, W.L.; Hemrika, W.; De Groot, P.G. Platelets express three different splice variants of ApoER2 that are all involved in signaling. *J. Thromb. Haemost.* **2007**, *5*, 1538–1544. [CrossRef] [PubMed]

98. Kolyada, A.; Porter, A.; Beglova, N. Inhibition of thrombotic properties of persistent autoimmune anti-β2GPI antibodies in the mouse model of antiphospholipid syndrome. *Blood* **2014**, *123*, 1090–1097. [CrossRef] [PubMed]

99. Palatinus, A.A.; Ahuja, K.D.; Adams, M.J. Effects of antiphospholipid antibodies on *in vitro* platelet aggregation. *Clin. Appl. Thromb. Hemost.* **2012**, *18*, 59–65. [CrossRef] [PubMed]

100. Betts, N.A.; Ahuja, K.D.; Adams, M.J. Anti-β2GP1 antibodies have variable effects on platelet aggregation. *Pathol.-J. RCPA* **2013**, *45*, 155–161. [CrossRef] [PubMed]

101. Sikara, M.P.; Routsias, J.G.; Samiotaki, M.; Panayotou, G.; Moutsopoulos, H.M.; Vlachoyiannopoulos, P.G. β2 Glycoprotein I (β2GPI) binds platelet factor 4 (PF4): Implications for the pathogenesis of antiphospholipid syndrome. *Blood* **2010**, *115*, 713–723. [CrossRef] [PubMed]

102. Ashby, B.; Daniel, J.L.; Smith, J.B. Mechanisms of platelet activation and inhibition. *Hematol. Oncol. Clin. North. Am.* **1990**, *4*, 1–26. [PubMed]

103. Li, Z.; Delaney, M.K.; O'Brien, K.A.; Du, X. Signaling during platelet adhesion and activation. *Arterioscler. Thromb. Vasc. Biol.* **2010**, *30*, 2341–2349. [CrossRef] [PubMed]

104. Pennings, M.; Derksen, R.; Van Lummel, M.; Adelmeijer, J.; Vanhoorelbeke, K.; Urbanus, R.; Lisman, T.; De Groot, P. Platelet adhesion to dimeric β_2-glycoprotein I under conditions of flow is mediated by at least two receptors: Glycoprotein Ibα and apolipoprotein E receptor 2'. *J. Thromb. Haemost.* **2007**, *5*, 369–377. [CrossRef] [PubMed]

105. Lutters, B.C.; Derksen, R.H.; Tekelenburg, W.L.; Lenting, P.J.; Arnout, J.; De Groot, P.G. Dimers of β_2-glycoprotein I increase platelet deposition to collagen via interaction with phospholipids and the apolipoprotein E receptor 2'. *J. Biol. Chem.* **2003**, *278*, 33831–33838. [CrossRef] [PubMed]

106. Mesquita, H.L.D.; Carvalho, G.R.D.; Aarestrup, F.M.; Correa, J.O.D.A.; Azevedo, M.R.A. Evaluation of platelet aggregation in the presence of antiphospholipid antibodies: Anti-β2GP1 and anticardiolipin. *Rev. Bras. Reumatol.* **2013**, *53*, 400–404. [CrossRef]

107. Ho, Y.C.; Ahuja, K.D.; Adams, M.J. Effects of anti-β2GP1 antibodies on collagen induced platelet aggregation. 2016, in preparation.

antibodies

MDPI

Article

A Clinical Approach for Defining the Threshold between Low and Medium Anti-Cardiolipin Antibody Levels for QUANTA Flash Assays

Gabriella Lakos *, Chelsea Bentow and Michael Mahler

Inova Diagnostics, Inc., 9900 Old Grove Road, San Diego, CA 92131-1638, USA;
cbentow@inovadx.com (C.B.); mmahler@inovadx.com (M.M.)
* Correspondence: glakos@inovadx.com; Tel.: +1-858-586-9900; Fax: +1-858-586-1401

Academic Editor: Ricard Cervera
Received: 14 March 2016; Accepted: 16 May 2016; Published: 25 May 2016

Abstract: The threshold between low and medium antibody levels for anticardiolipin (aCL) and anti-β2 glycoprotein I antibodies (aβ2GPI) for the diagnosis of antiphospholipid syndrome (APS) remains a matter of discussion. Our goal was to create a protocol for determining the low/medium antibody cut-off for aCL antibody methods based on a clinical approach, and utilize it to establish the clinically-relevant low/medium threshold for QUANTA Flash aCL chemiluminescent immunoassay (CIA) results. The study included 288 samples from patients with primary APS ($n = 70$), secondary APS ($n = 42$), suspected APS ($n = 36$), systemic lupus erythematosus (SLE) without APS ($n = 96$) and other connective tissue diseases ($n = 44$). All samples were tested for IgG and IgM aCL antibodies with QUANTA Flash CIA, along with traditional enzyme-linked immunosorbent assays (ELISAs) (QUANTA Lite). The assay specific low/medium threshold for QUANTA Flash aCL IgG and IgM assays (*i.e.*, the equivalent of 40 GPL and MPL units) was established as 95 and 31 chemiluminescent units (CU), respectively, based on clinical performance and comparison to QUANTA Lite ELISAs. Agreement between CIA and ELISA assay results improved substantially when the platform-specific low/medium antibody threshold was used, as compared to agreement obtained on results generated with the assay cutoff: Cohen's *kappa* increased from 0.85 to 0.91 for IgG aCL, and from 0.59 to 0.75 for IgM aCL results. This study describes a clinical approach for establishing the low/medium antibody threshold for aPL antibody assays, and successfully employs it to define 95 and 31 CU, respectively, as the low/medium cut point for QUANTA Flash aCL IgG and IgM results. This study can serve as a model for labs wishing to establish the appropriate low/medium aPL antibody threshold when implementing new aPL antibody assays.

Keywords: antiphospholipid syndrome; anticardiolipin antibodies; low/medium antibody threshold; chemiluminescent immunoassay

1. Introduction

The updated classification criteria for definite antiphospholipid syndrome (APS), also known as Hughes syndrome, specifies anticardiolipin (aCL) and anti-β_2-glycoprotein I (β_2GPI) antibodies of IgG and/or IgM isotype in medium or high titer as one of the laboratory criteria [1]. As inter-laboratory agreement between aCL measurements is known to be poor due to inconsistencies of the cut-off, calibration, and other methodological issues [2–4], the committee recommends reporting positive results in ranges of positivity (*i.e.*, low-medium-high) to achieve better inter-run and inter-laboratory agreement than that obtained with quantitative results only [1,5]. For aCL antibodies measured by enzyme-linked immunosorbent assays (ELISA), the international consensus states that values above 40 IgG and IgM Phospholipid (GPL or MPL) units, or above the 99th percentile of the values obtained on reference subjects are considered medium or high titer aCL antibodies. The committee overseeing the revised classification criteria has acknowledged the lack of suitable evidence on this issue, but stated that these values should be used "until international consensus is reached" [1].

This concept, however, has several shortcomings. First, the 99th percentile often defines values which are significantly different from the recommended 40 GPL or MPL units [5]. In fact, the value depends on the performance characteristics of the particular assay, the statistical method, and the reference population that is used to establish the cut-off. Additionally, in the absence of a reference method, and in the light of the analytical diversity of aPL antibody assays, the use of the same unit type (GPL and MPL) by itself is not sufficient to achieve harmonization between antiphospholipid (aPL) antibody assays. This is evident by the different cut-off values of different brands of kits, and the wide range of results reported by labs during proficiency testing surveys [2–5]. Differences exist not only between various traditional, ELISA-based tests, but also between traditional tests and new technologies, such as chemiluminescent immunoassays (CIA) and addressable laser bead immunoassays (ALBIA) [6–8]. The analytical performance characteristics of these tests are often different from that of traditional technologies. Therefore, using the same low/medium threshold for all assays is unlikely to be an optimal to approach to achieve consistent and appropriate patient management.

To be able to leverage laboratory automation, aPL assays are being increasingly replaced with newer assays in the clinical lab. The switch from one method to another may be challenging for aPL antibodies, and if the change means the introduction of a different unit type, cut-off or analytical measuring range, it may create interpretation challenges. To prevent unfavorable effects on patient care, new methods should be carefully evaluated, compared to the traditional methods, and potential differences in unit values, unit types, and low-medium-high categories need to be analyzed and properly interpreted.

Our goal was to create and employ a protocol for the establishment of the clinically-relevant (low/medium) threshold for QUANTA Flash aCL IgG and IgM microparticle chemiluminescent immunoassays. Following the 14th International Congress on Antiphospholipid Antibodies, a committee of experts in the field of APS proposed that the threshold for aPL antibody levels should be determined using clinical approach [9], specifically, by considering the performance of a particular assay for the association with APS-related clinical symptoms. Therefore, we have set out to determine the low/medium cut-off for the QUANTA Flash aCL IgG, and IgM methods based on the clinical performance of these new tests, using traditional ELISA as reference.

2. Results

2.1. Analytical Performance

To verify the analytical performance of the QUANTA Flash IgG and IgM aCL assays, precision and linearity studies were performed. The within-run coefficients of variation (%CV) for the high and the low controls of the QUANTA Flash assays ranged from 1.0% to 3.4%. The between-day %CV ranged from 1.2% to 6.0%, and the total imprecision was between 1.5% and 6.2%. For the linearity study, results obtained on two serially-diluted samples per assay were combined in one linear regression plot. The slopes of the regression lines were 0.98 and 0.99, respectively, with coefficient of determinations (R^2) of 1.00.

2.2. Threshold between Low and Medium Antibody Levels

40 GPL and MPL have previously been suggested as the thresholds between low and medium-high aCL antibody levels. To verify the relevance of this value as a clinically significant antibody titer, we determined the clinical sensitivity and specificity of the QUANTA Lite aCL ELISA assays for APS-related clinical symptoms (venous thrombosis, arterial thrombosis, and obstetric complications) at the 40 GPL and MPL level. At this threshold, the sensitivity of the aCL IgG and IgM ELISA was 48.1% and 25.0%, with 91.0% and 92.4% specificity, respectively (Table 1). These values indicate that at 40 GPL

and MPL cut-off, the aCL ELISAs indeed deliver clinically-relevant results. Next, we performed receiver operating characteristic (ROC) analysis on QUANTA Flash aCL IgG, and IgM results to calculate the threshold that provides the same or similar clinical performance (sensitivity and specificity) (Figure 1). We were able to identify CU thresholds where the clinical sensitivity of the QUANTA Flash aCL IgG and IgM assays was essentially identical to that of the QUANTA Lite tests at the 40 unit threshold. The associated specificity values were also the same as those for QUANTA Lite. These threshold values were determined to be 31 CU for aCL IgM and 95 CU for aCL IgG (Table 1). These data points can be identified on the ROC curves as a point where the two curves cross each other (Figure 1). These results indicate that QUANTA Flash aCL assays deliver similar clinical performance at 95 and 31 CU threshold (for IgG and IgM, respectively) as that of the QUANTA Lite assays at the 40 GPL and MPL cut-off; in other words, the results suggest the equivalency of the QUANTA Flash 95 and 31 CU with the conventional 40 GPL and MPL low/medium threshold commonly utilized in traditional assays.

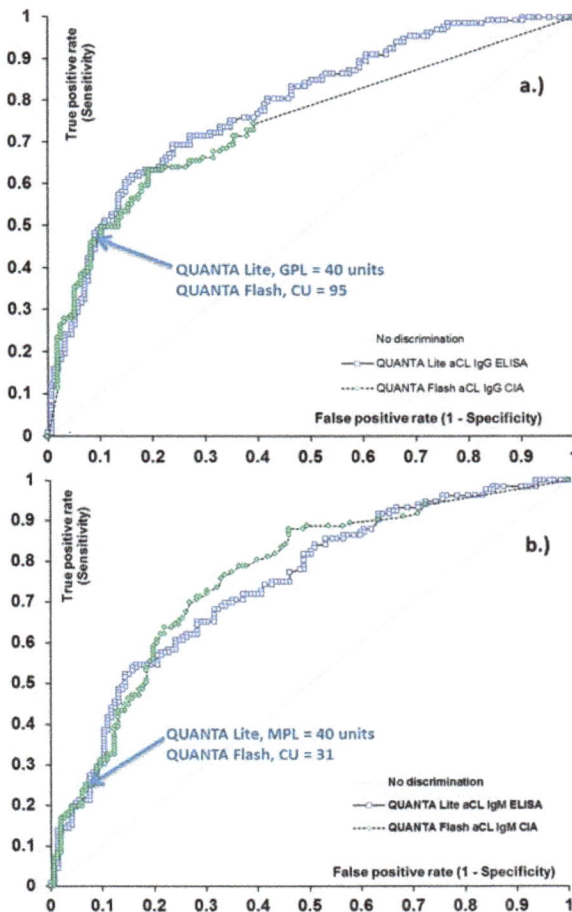

Figure 1. Receiver operating characteristic (ROC) analysis of aCL IgG (**a**) and IgM (**b**) methods for antiphospholipid syndrome (APS)-related clinical symptoms. Arrows indicate the clinically relevant threshold between low and medium titer for aCL assays.

Table 1. Low/medium threshold values, and associated clinical sensitivity and specificity of anticardiolipin (aCL) assays. GPL: IgG Phospholipid; CU: chemiluminescent units; MPL: IgM Phospholipid; CI: confidence interval.

Assay Characteristic	QUANTA Lite aCL IgG	QUANTA Flash aCL IgG	QUANTA Lite aCL IgM	QUANTA Flash aCL IgM
Low/Medium Threshold Unit	40 GPL	95 CU	40 MPL	31 CU
Sensitivity, % (95% CI) at Threshold	48.1 (39.4–56.9)	48.1 (39.4–56.9)	25.0 (17.9–33.3)	25.0 (17.9–33.3)
Specificity, % (95% CI) at Threshold	91.0 (85.3–95.0)	89.7 (83.8–94.0)	92.4 (86.8–96.2)	92.4 (86.8–96.2)

2.3. Qualitative Agreement and Quantitative Correlation between Methods

ROC curve analysis resulted in area under the curve (AUC) values ranging from 0.72 to 0.78 for ELISA and CIA methods, and demonstrated very similar diagnostic performance for the two platforms (Figure 1). Additionally, good qualitative agreement was found between CIA and ELISA methods, with overall agreements ranging from 84.9% (aCL IgM assays) to 93.1% (aCL IgG assays). Cohen's kappa coefficients were 0.59 and 0.85, implying moderate to substantial agreement. Quantitative results also showed significant correlation between the methods, with Spearman's rho of 0.74 and 0.83 ($p < 0.0001$ for both) (Table 2). The analysis of discrepant results revealed that the majority of these samples had low positive (*i.e.*, clinically less significant) antibody levels. Indeed, when the platform-specific threshold between low and medium positive samples (40 GPL and 40 MPL units for aCL ELISAs, and 95 CU and 31 CU for QUANTA Flash aCL IgG and IgM assays) was used as the cut-off, the agreement between the platforms substantially improved, with total agreement reaching 96.5% and kappa of 0.91 for aCL IgG assays, and total agreement of 93.5% and kappa of 0.75 for aCL IgM assays (Table 2).

Table 2. Qualitative agreement and quantitative correlation between chemiluminescent immunoassay (CIA) and enzyme-linked immunosorbent assay (ELISA) methods at the assay-specific cut-off values and at assay-specific low/medium threshold.

		QUANTA Flash CIA *vs.* QUANTA Lite ELISA		
		Assay	IgG	IgM
aCL	at the cut-off	Total % agreement (95% CI)	93.1 (89.5–95.7)	84.9 (80.1–88.9)
		Cohen's kappa coefficient (95% CI)	0.85 (0.78–0.91)	0.59 (0.48–0.70)
		Spearman's rho (*p*)	0.83 ($p < 0.0001$)	0.74 ($p < 0.0001$)
	At low/medium threshold	Total % agreement (95% CI)	96.5 (93.7–98.3)	93.5 (89.9–96.1)
		Cohen's kappa coefficient (95% CI)	0.91 (0.86–0.97)	0.75 (0.64–0.86)

3. Discussion

This study describes an experimental protocol for determining the low/medium antibody threshold for aPL antibody methods. Using this approach, we have identified 95 and 31 CU as low/medium threshold for results generated with the QUANTA Flash aCL IgG and IgM assays, respectively.

Although the association between thrombotic complications and antiphospholipid antibodies was first demonstrated more than 30 years ago [10], APS still poses diagnostic challenges in routine clinical practice. The characteristic clinical symptoms of the disease are actually more frequently present in non-APS than in APS patients, and the hallmark antibodies of the syndrome can occur as natural or infection-induced antibodies [1]. Rigorous specification of the clinical symptoms and laboratory results promotes accurate diagnosis [1]; however, the analytical diversity and less than optimal reproducibility of aPL results continue to make interpretation of aPL antibody results challenging. Defining the threshold between low (*i.e.*, clinically less significant) and medium-high (clinically more significant) aCL and β2GPI antibody levels helps distinguish APS patients from other diseases, but may lead to inappropriate decisions if interpreted improperly. In spite of continuous harmonization efforts [11,12], inter-laboratory portability of aPL results remains suboptimal [2–4], and the emergence of new platforms and technologies are bringing additional analytical variability and potential confusion into the measurement process.

The QUANTA Flash aCL and β2GPI tests are microparticle-based chemiluminescent immunoassays. Although the clinical performance of these assays has been found to be good [13–16], the wide AMR and the use of arbitrary chemiluminescent units have created challenges about the interpretation of the numerical unit values [14,17]. In particular, the lack of definition for low/medium antibody threshold hinders diagnostic efforts. Mathematical conversion of CU values to GPL and MPL units was found to be impractical, as the CIA and ELISA assays have very different analytical and technological characteristics. Although both platforms have the same numerical cut-off (20 CU for QUANTA Flash assays and 20 GPL or MPL units for QUANTA Lite assays), the correlation between unit values is non-linear, due to the wider AMR, and the better resolution and dilution linearity of QUANTA Flash results [18]. In this study we have chosen to approach the problem from a clinical point of view, as recommended by the 14th International Congress on Antiphospholipid Antibodies Task Force [9].

We verified the clinical relevance of the 40 GPL and MPL unit threshold for the QUANTA Lite aCL assays, and determined that, at this level, the ELISAs were are able to distinguish APS patients from non-APS patients with acceptable clinical sensitivity (48.1% and 25.0%) and specificity (91% and 92.4%). Based on this performance goal, we utilized ROC analysis to identify 95 CU for IgG and 31 CU for IgM aCL antibodies as thresholds for QUANTA Flash assays providing equivalent clinical performance as that of the ELISAs at 40 GPL and MPL units. These values are, therefore, considered as the low/medium threshold for aCL antibodies measured with QUANTA Flash tests.

QUANTA Flash and QUANTA Lite aCL results showed moderate to substantial qualitative agreement (84.9% and 93.1% for IgM and IgG, respectively), and significant quantitative correlation (Spearman rho 0.74 to 0.86, $p < 0.0001$). Agreement improved significantly to 93.5% and 96.5% for IgM and IgG aCL, respectively, when the assay-specific low/medium threshold was used as the cut-off.

Establishing the low/medium antibody threshold for QUANTA Flash aCL antibody results will facilitate the utilization and help achieve correct interpretation of the results. It also ensures the continuity and consistency of patient care by using low/medium cut points that are clinically equivalent to those described in the classification criteria. In addition, as the study protocol can be utilized for any new aPL antibody test, this study can serve as a model for labs wishing to establish the appropriate low/medium aPL antibody threshold when implementing new aPL antibody assays.

4. Materials and Methods

4.1. Samples

The study included 288 samples collected at the Clinical Division of Allergy and Immunology at Jagiellonian University Medical College (Krakow, Poland) from patients referred to the clinic with the diagnosis of SLE, other systemic autoimmune disease and/or APS. The population comprised of samples from patients with primary APS ($n = 70$), secondary APS ($n = 42$) (all SLE), suspected APS patients ($n = 36$), and control sera from patients with SLE without APS ($n = 96$) and other connective tissue diseases ($n = 44$, Sjogren's, syndrome, dermatomyositis, mixed connective tissue disease, scleroderma, undifferentiated connective tissue disease). Suspected APS patients did not completely fulfill the classification criteria, but were either aPL antibody-positive without classical clinical symptoms, or had one of the criteria clinical symptoms without medium or high levels of aPL antibody positivity. Data on the presence or absence of venous thrombosis, arterial thrombosis, and obstetric complications were available for all patients. APS diagnosis was made based on the updated Sydney APS criteria [1]. SLE patients were diagnosed according to the American College of Rheumatology criteria whenever at least four ACR criteria were fulfilled [19]. All other diagnoses were established as described before [20]. This study meets and is in compliance with all ethical standards in medicine, and informed consent was obtained from all patients according to the Declaration of Helsinki. All samples were tested for aCL antibodies using QUANTA Flash® aCL (IgG, IgM) and QUANTA Lite aCL (IgG, IgM).

4.2. QUANTA Flash® Methods

The QUANTA Flash aCL (IgG and IgM) assays (Inova Diagnostics Inc., San Diego, CA, USA) are microparticle chemiluminescent immunoassays (CIAs) that are run on the BIO-FLASH® instrument (Biokit S.A., Barcelona, Spain). BIO-FLASH is a random access, rapid-response, fully-automated chemiluminescent analyzer. Results are expressed in (arbitrary) chemiluminescent units (CU). Analytical characteristics of the assays, including cut-off, measure of units and analytical measuring range (AMR) are summarized in Table 3.

Table 3. Analytical characteristics of the aCL antibody assays used in this study.

Assay	Antigen	Units of Measurement	Analytical Measuring Range	Cut-Off Value (Reference Ranges)
QF aCL IgG	Cardiolipin and human β2GPI	CU	2.6–2024 CU	⩾20 Positive
QF aCL IgM	Cardiolipin and human β2GPI	CU	1.0–774 CU	⩾20 Positive
QL aCL IgG	Cardiolipin and bovine β2GPI	GPL	0.0–150.0 GPL	<15 Negative 15–20 Indeterminate >20 Positive
QL aCL IgM	Cardiolipin and bovine β2GPI	MPL	0.0–150.0 MPL	<12.5 Negative 12.5–20 Indeterminate >20 Positive

QF = QUANTA Flash; QL = QUANTA Lite; CU = chemiluminescent units.

4.3. QUANTA Lite® Methods

The QUANTA Lite aCL (IgG and IgM) methods (Inova Diagnostics, San Diego, CA, USA) are traditional enzyme-linked immunosorbent assays (ELISAs) for the semi-quantitative determination of aCL antibodies in human serum. The QUANTA Lite aCL assays report results in GPL and MPL units. All QUANTA Lite ELISAs were performed according to the manufacturer's guidelines. Analytical characteristics of the assays are summarized in Table 3. QUANTA Lite aCL assays have an equivocal range defined. For the purposes of this study, only values above the equivocal range were defined as positive.

4.4. Analytical Performance Assessment of QUANTA Flash Methods

Precision performance and linearity of the QUANTA Flash aCL assays were verified as part of the analytical assessment. Testing was performed according to relevant Clinical and Laboratory Standards Institute (CLSI) guidelines EP5-A2 and EP6-A. Within-run, between-days and total imprecision were determined by running two samples (the low and the high controls) in triplicate for five days. Linearity testing was performed by serially diluting two samples (one high and one low) to span the AMR for each assay, testing the dilutions in duplicate, plotting obtained values against expected values, and analyzing the results with linear regression.

4.5. Statistical Analyses

Data were statistically evaluated using the Analyse-it for Excel software (Version 2.30; Analyse-it Software, Ltd., Leeds, UK). Cohen's *kappa* agreement test was used to assess concordance between portions, and Spearman's correlation was used to evaluate quantitative relationship between unit values. Outcome was considered significant if *p* value was less than 0.05. Receiver operating characteristics (ROC) analysis was used to assess the diagnostic performance of the different immunoassays.

5. Conclusions

As recommended by the international committee of experts in APS, this study uses a clinical approach for establishing the low/medium antibody threshold for QUANTA Flash aCL IgG and IgM methods. This analysis will help achieve the correct interpretation of the results; moreover, it can serve as a model for labs wishing to establish the appropriate low/medium aPL antibody threshold when implementing new aPL antibody assays.

Acknowledgments: We are grateful to J. Musiał and T. Iwaniec (Clinical Division of Allergy and Immunology, Jagiellonian University Medical College, Krakow, Poland) for providing patient samples and performing the aPL antibody testing.

Author Contributions: G.L. and C.B. compiled the data and performed statistical analysis for the paper. G.L., C.B. and M.M. wrote the paper.

Conflicts Disclosure Statement: C.B., G.L. and M.M. are employees of Inova Diagnostics.

References

1. Miyakis, S.; Lockshin, M.D.; Atsumi, T.; Branch, D.W.; Brey, R.L.; Cervera, R.; Derksen, R.H.; DE Groot, P.G.; Koike, T.; Meroni, P.L.; *et al.* International consensus statement on an update of the classification criteria for definite antiphospholipid syndrome (APS). *J. Thromb. Haemost.* **2006**, *4*, 295–306. [CrossRef] [PubMed]
2. Favaloro, E.J.; Silvestrini, R. Assessing the usefulness of anticardiolipin antibody assays: a cautious approach is suggested by high variation and limited consensus in multilaboratory testing. *Am. J. Clin. Pathol.* **2002**, *118*, 548–557. [CrossRef] [PubMed]
3. Wong, R.; Favaloro, E.; Pollock, W.; Wilson, R.; Hendle, M.; Adelstein, S.; Baumgart, K.; Homes, P.; Smith, S.; Steele, R.; *et al.* A multi-centre evaluation of the intra-assay and inter-assay variation of commercial and in-house anti-cardiolipin antibody assays. *Pathology* **2004**, *36*, 182–192. [CrossRef] [PubMed]
4. Reber, G.; Tincani, A.; Sanmarco, M.; de Moerloose, P.; Boffa, M.C. Variability of anti-beta2 glycoprotein I antibodies measurement by commercial assays. *Thromb. Haemost.* **2005**, *94*, 665–672. [PubMed]
5. Ruffatti, A.; Olivieri, S.; Tonello, M.; Bortolati, M.; Bison, E.; Salvan, E.; Facchinetti, M.; Pengo, V. Influence of different IgG anticardiolipin antibody cut-off values on antiphospholipid syndrome classification. *J. Thromb. Haemost.* **2008**, *6*, 1693–1696. [CrossRef] [PubMed]
6. Montaruli, B.; De Luna, E.; Erroi, L.; Marchese, C.; Mengozzi, G.; Napoli, P.; Nicolo', C.; Romito, A.; Bertero, M.T.; Sivera, P.; *et al.* Analytical and clinical comparison of different immunoassay systems for the detection of antiphospholipid antibodies. *Int. J. Lab. Hematol.* **2016**, *38*, 172–182. [CrossRef] [PubMed]
7. Capozzi, A.; Lococo, E.; Grasso, M.; Longo, A.; Garofalo, T.; Misasi, R.; Sorice, M. Detection of antiphospholipid antibodies by automated chemiluminescence assay. *J. Immunol. Methods* **2012**, *379*, 48–52. [CrossRef] [PubMed]
8. Persijn, L.; Decavele, A.S.; Schouwers, S.; Devreese, K. Evaluation of a new set of automated chemiluminescense assays for anticardiolipin and anti-beta2-glycoprotein I antibodies in the laboratory diagnosis of the antiphospholipid syndrome. *Thromb. Res.* **2011**, *128*, 565–569. [CrossRef] [PubMed]
9. Bertolaccini, M.L.; Amengual, O.; Andreoli, L.; Atsumi, T.; Chighizola, C.B.; Forastiero, R.; Lakos, G.; Lambert, M.; Meroni, P.; Ortel, T.L.; *et al.* 14th International Congress on Antiphospholipid Antibodies Task Force. Report on antiphospholipid syndrome laboratory diagnostics and trends. *Autoimmun. Rev.* **2014**, *13*, 917–930. [CrossRef] [PubMed]

10. Hughes, G.R. Hughes syndrome/APS. 30 years on, what have we learnt? Opening talk at the 14th International Congress on antiphospholipid antibodies Rio de Janiero, October 2013. *Lupus* **2014**, *23*, 400–406. [CrossRef] [PubMed]

11. Willis, R.; Lakos, G.; Harris, E.N. Standardization of antiphospholipid antibody testing—Historical perspectives and ongoing initiatives. *Semin. Thromb. Hemost.* **2014**, *40*, 172–177. [PubMed]

12. Willis, R.; Harris, E.N.; Pierangeli, S.S. Current international initiatives in antiphospholipid antibody testing. *Semin. Thromb. Hemost.* **2012**, *38*, 360–374. [CrossRef] [PubMed]

13. Zhang, S.; Wu, Z.; Li, P.; Bai, Y.; Zhang, F.; Li, Y. Evaluation of the Clinical Performance of a Novel Chemiluminescent Immunoassay for Detection of Anticardiolipin and Anti-Beta2-Glycoprotein 1 Antibodies in the Diagnosis of Antiphospholipid Syndrome. *Medicine (Baltim.)* **2015**, *94*. [CrossRef] [PubMed]

14. Van Hoecke, F.; Persijn, L.; Decavele, A.S.; Devreese, K. Performance of two new, automated chemiluminescence assay panels for anticardiolipin and anti-beta2-glycoprotein I antibodies in the laboratory diagnosis of the antiphospholipid syndrome. *Int. J. Lab. Hematol.* **2012**, *34*, 630–640. [CrossRef] [PubMed]

15. Meneghel, L.; Ruffatti, A.; Gavasso, S.; Tonello, M.; Mattia, E.; Spiezia, L.; Campello, E.; Hoxha, A.; Fedrigo, M.; Punzi, L.; *et al.* The clinical performance of a chemiluminescent immunoassay in detecting anti-cardiolipin and anti-beta2 glycoprotein I antibodies. A comparison with a homemade ELISA method. *Clin. Chem. Lab. Med.* **2015**, *53*, 1083–1089. [CrossRef] [PubMed]

16. Iwaniec, T.; Kaczor, M.P.; Celinska-Lowenhoff, M.; Polanski, S.; Musial, J. Identification of patients with triple antiphospholipid antibody positivity is platform and method independent. *Polskie Arch. Med. Wewn.* **2016**, *126*, 19–24. [CrossRef]

17. Devreese, K.M.; Van, H.F. Anticardiolipin and anti-beta2glycoprotein-I antibody cut-off values in the diagnosis of antiphospholipid syndrome: More than calculating the in-house 99th percentiles, even for new automated assays. *Thromb. Res.* **2011**, *128*, 598–600. [CrossRef] [PubMed]

18. Lakos, G. Analytical Detection Capabilities of Immunoassay-Based Antiphospholipid Antibody Tests: Do They Matter? *Drug Dev. Res.* **2013**, *74*, 575–581. [CrossRef]

19. Tan, E.M.; Cohen, A.S.; Fries, J.F.; Masi, A.T.; McShane, D.J.; Rothfield, N.F.; Schaller, J.G.; Talal, N.; Winchester, R.J. The 1982 revised criteria for the classification of systemic lupus erythematosus. *Arthritis Rheum.* **1982**, *25*, 1271–1277. [CrossRef] [PubMed]

20. Mahler, M.; Fritzler, M.J.; Bluthner, M. Identification of a SmD3 epitope with a single symmetrical dimethylation of an arginine residue as a specific target of a subpopulation of anti-Sm antibodies. *Arthritis Res. Ther.* **2005**, *7*, R19–R29. [CrossRef] [PubMed]

antibodies

MDPI

Review
Antiphospholipid Antibodies: Their Origin and Development

Karl J. Lackner * and Nadine Müller-Calleja

Institute of Clinical Chemistry and Laboratory Medicine, University Medical Center Mainz, D-55101 Mainz, Germany; nadine.prinz@unimedizin-mainz.de
* Correspondence: karl.lackner@unimedizin-mainz.de; Tel.: +49-613-117-7190

Academic Editor: Ricard Cervera
Received: 28 April 2016; Accepted: 19 May 2016; Published: 2 June 2016

Abstract: Antiphospholipid antibodies (aPL) are a hallmark of the antiphospholipid syndrome (APS), which is the most commonly acquired thrombophilia. To date there is consensus that aPL cause the clinical manifestations of this potentially devastating disorder. However, there is good evidence that not all aPL are pathogenic. For instance, aPL associated with syphilis show no association with the manifestations of APS. While there has been intensive research on the pathogenetic role of aPL, comparably little is known about the origin and development of aPL. This review will summarize the current knowledge and understanding of the origin and development of aPL derived from animal and human studies.

Keywords: antiphospholipid antibodies; natural antibodies; innate immunity; B1 B cells

1. Introduction

Antibodies against phospholipids have been known for many decades as a hallmark of infection with *Treponema pallidum*. In 1906, Wassermann introduced a complement binding assay to detect antibodies in syphilitic patients [1]. Landsteiner soon hypothesized that the antigen might be a lipid rather than a protein [2], but it took over three decades until it was shown that the antigen in this assay was a phospholipid. This lipid was later called cardiolipin, because it was purified from myocardium [3]. With the continued use of cardiolipin based serologic assays for the diagnosis of syphilis it became apparent that a small group of patients with autoimmune disease, mostly systemic lupus erythematosus (SLE) had "false positive" tests caused by autoantibodies against cardiolipin. In the 1980s, researchers recognized that the presence of so called antiphospholipid antibodies (aPL) in SLE patients was associated with thromboembolic events and recurrent abortions, and the term anticardiolipin syndrome and later antiphospholipid syndrome (APS) was coined [4,5].

Today, there is broad consensus that aPL cause the clinical manifestations of APS. However, the underlying mechanisms are still a matter of controversy. This is perhaps related to the broad heterogeneity of aPL. Some aPL bind to anionic or neutral phospholipids coated to microtiter plates in the absence of proteins. Others can only bind in the presence of specific protein cofactors, e.g., β2-glycoprotein I (β2GPI) or prothrombin. The latter aPL are called cofactor dependent. Yet another group of aPL binds to the cofactors. These are also regarded as aPL even though their antigens in the strict sense are proteins or peptides. Some of the aPL detected by immunoassays also inhibit phospholipid dependent clotting assays. These are collectively called lupus anticoagulants (LA) [6]. It should be noted that there are some LA which do not react in the traditional immunoassays.

While there has been tremendous progress in the understanding of the pathogenic potential of aPL which has been reviewed repeatedly in the recent past [7–12], relatively little is known about the origin of aPL. As mentioned above patients suffering from syphilis develop antibodies against cardiolipin during their infection. However, these aPL do not induce the clinical symptoms of APS and must be regarded as different from pathogenic aPL. Similarly, it has been shown that other infectious diseases may cause the transient appearance of aPL. Again it appears that these transient aPL do not contribute to the development of APS. However, it has never been excluded that these transient antibodies might be pathogenic, but do not cause relevant damage, because of their transient nature. And finally, there have been reports of patients with monoclonal gammopathy with a monoclonal aPL. Interestingly, no such patient has been described with the clinical picture of APS. Until now there has been no scientifically proven explanation why some patients develop pathogenic aPL and subsequently APS. We will review the current knowledge about the origin and maturation of aPL and try to put forward some hypotheses on the development of pathogenic aPL.

2. Are aPL Part of the Natural Antibody Repertoire?

Natural antibodies appear without prior infection or immunization. The majority is of the immunoglobulin (Ig) M isotype, but IgG or IgA have also been observed [13,14]. They are secreted mainly by B1 cells, a specific subset of B-lymphocytes. Activation of B1 cells does not depend on antigenic challenge and T-cell help, but can be elicited by constituents of innate immunity, e.g., pathogen associated molecular patterns (PAMPs). Natural antibodies are usually of low to moderate affinity but cross-react with several related antigens including autoantigens. Sequence analysis shows that natural antibodies are usually very close to germline sequences with few if any somatic mutations. It is postulated that natural antibodies constitute a rapid, first line response to infection that bridges the time needed by adaptive immunity to develop specific antibodies. An example are antibodies to phosphorylcholine, a constituent of Gram positive cell walls. Lack of B1 cells severely compromises the resistance to bacterial infections. Interestingly, also antibodies against phosphatidylcholine have been identified which is a component of senescent cell membranes. This suggests that natural antibodies also play a role in the removal of dying cells. This function in removal of possibly antigenic debris might also explain the protection against autoimmunity conferred by B1 cells.

It has been proposed in the past that aPL belong to the natural antibodies [15,16], because they share many properties with these B1 cell derived antibodies. aPL tend to be polyspecific and there is overlap with other autoantibodies e.g., anti-DNA. Many aPL are germline encoded or exhibit only minor deviations from germline sequences (see below). However, final proof of this concept has not been provided. We will review the current evidence that aPL belong to the natural antibody repertoire and that even germline encoded aPL may be pathogenic.

2.1. Animal Models

Animal models permit a more detailed analysis of the mechanisms how aPL develop. Unfortunately, also in the mouse model, data are by no means conclusive. However, there is an interesting mouse model of APS which strongly supports the notion that aPL are natural antibodies. This model is based on immunization of animals with an aPL in the presence of an appropriate adjuvant. Mice immunized in this way develop their own aPL. This model has been described by the group of Yehuda Shoenfeld in the early 1990s and has been used by other researchers as a model of APS [17]. It was initially explained by the generation of anti-idiotypic antibodies. Immunization with an aPL was proposed to generate an antibody against this specific aPL. This was supposedly followed by generation of an anti-idiotype that would have similar binding specificity as the original aPL used for immunization [18]. This concept has never been proven experimentally. The time course of the antibody response makes this sequence of events highly unlikely. Pierangeli and colleagues [19] showed that immunized animals develop very rapidly, *i.e.*, within one week after the first immunization, their own aPL reactive against cardiolipin while no anti-β2GPI is induced in this time frame. Furthermore, most of these aPL are of the IgG and not the IgM isotype. Considering the antigen used for immunization, the time frame in which aPL occur, and the fact that most aPL produced in this model are of the IgG isotype, the usual response of the adaptive immune system to an antigenic challenge cannot account for this phenomenon. First, the antigen used for immunization has nothing to do with the immediate antibody response. Second, the adaptive immune system does not usually generate significant amounts of specific IgG antibodies within 1 week. Thus, it is highly likely that this immunization scheme somehow induces a natural antibody response. Apparently, most of the antibody produced is of the IgG-type which is unusual but clearly possible for natural antibodies. Most importantly, aPL induced by this protocol have been shown to be pathogenic *in vivo*. Immunized mice develop thrombophilia as well as pregnancy failure [17,20]. Thus, in summary we propose that this unique mouse model provides strong evidence that aPL of the IgG isotype belong to the natural antibody repertoire and that these aPL are pathogenic, at least in mice. Furthermore, the rapid induction of pathogenic aPL implies that antigen driven maturation is not an absolute requirement for pathogenicity.

Along these lines it may be relevant that aPL have been shown to sensitize antigen presenting cells including plasmacytoid dendritic cells (pDC) towards agonists of toll-like receptor (TLR) 7 and/or TLR8 [21]. As a consequence exposure to single stranded RNA (ssRNA) or other agonists leads to a massively increased secretion of type I interferon. Unbalanced activation of TLR7 in particular in pDC has been shown to induce autoimmunity and autoantibody production in mice [22–24]. Thus, the effects of aPL on pathways of innate immunity might help to better understand this mouse model of APS.

It should be noted that other induction schemes have been explored in mice and rabbits that also lead to pathogenic aPL. For instance, immunization of rabbits with lipid A can also rapidly induce pathogenic aPL [25].

2.2. Infections and aPL

In humans, many infectious diseases are associated with a transient or permanent rise of aPL of the IgM and IgG isotype. These include viral infections, e.g., parvovirus B-19, cytomegalovirus (CMV) and hepatitis C, as well as bacterial and parasitic infections, e.g., syphilis or helicobacter pylori infection [26]. Even though the production of specific aPL by the adaptive immune system cannot be ruled out and molecular mimicry is proposed as one possible underlying mechanism [27–31], the high frequency of a uniform antibody response to extremely different antigens should alert to the possibility of induction of natural antibodies. Another interesting aspect of the association of viral infections with aPL is the fact that there is a significant number of patients who develop thrombotic events [26,32–34]. While it is not proven that these events are caused by aPL, the undisputable coincidence raises the question whether these infection associated aPL may be pathogenic.

2.3. Analysis of Human Monoclonal aPL

Analysis of monoclonal aPL isolated from patients with APS or healthy individuals has provided important insights into the natural history of aPL. Several monoclonal aPL including aPL of the IgG isotype show a germline configuration as would be expected if they belong to the natural antibody repertoire. A thorough review of the available sequence data on human monoclonal aPL has been published [35]. Overall, the data obtained from sequence analysis of human monoclonal aPL provide a heterogeneous picture. Some aPL have a germline sequence, but many aPL clearly show all signs of antigen driven maturation. While this suggests that these antibodies are derived from typical adaptive immune responses, the presence of somatic mutations does not rule out that the original antibody was part of the natural repertoire and produced by B1cells. In fact, isotype switches and somatic mutations in B1 cell derived antibodies occurs and has been discussed as an escape mechanism of autoimmune disease [36].

Along these lines the group of Jean-Louis Pasquali and co-workers could show in a series of elegant experiments that B cell clones producing IgG aPL are present in APS patients as well as in healthy individuals [37,38]. These B cell clones were surprisingly heterogeneous in terms of V-region usage. Furthermore, this group confirmed the presence of aPL with germline configuration as well as aPL with somatic mutations. Their data suggest that low affinity aPL belong to the natural antibody repertoire and that by so far unknown triggers these aPL can undergo antigen driven maturation [39].

Thus, there is a large body of evidence that many aPL including aPL of the IgG isotype are natural antibodies. However, it is also clear that antigen driven maturation of aPL and in particular anti-β2GPI does occur. It is not known, if this occurs starting from the natural antibodies or from completely different B-cell clones.

2.4. What Is the Role of Antigen Driven Maturation?

As outlined above several groups have isolated IgG aPL with significant deviations from known germline sequences. Thus, antigen driven maturation of aPL has been unequivocally proven. Some investigators have put forward the hypothesis that antigen driven maturation is required to generate pathogenic aPL. Lieby and colleagues isolated an aPL with three somatic mutations from an APS patient. This antibody was pathogenic in an *in vivo* pregnancy model. When this antibody was modified back to the germline sequence it still bound to phospholipids but was no longer pathogenic [40]. The authors interpreted this finding as indicating that pathogenicity of aPL is induced by antigen driven maturation and that this process is perhaps a prerequisite of pathogenicity.

We have also isolated human monoclonal aPL with germline configuration and somatic mutations [41–43]. Binding specificity of two of these antibodies, RR7F which has a high homology to germline and HL5B which carries several somatic mutations, is similar. Both induce potentially pathogenic responses in monocytes and endothelial cells, but the required concentration of RR7F is approx. one order of magnitude higher than that of HL5B [21,44,45]. However, both monoclonal aPL induce thrombus formation in an *in vivo* model of venous thrombosis [46]. Our data support the concept that antigen driven maturation does increase the pathogenic potential of aPL but that it is not an indispensable prerequisite for pathogenicity. With respect to the requirements for pathogenicity of aPL data of Girardi and co-workers [47] are of relevance. They confirmed previous data that the human monoclonal aPL Mab519 is pathogenic in mice. It causes foetal resorption in pregnant mice. Mab519 was cloned from a healthy individual and deviates only non-significantly from germline sequence [48].

In summary, antigen driven maturation is not required for pathogenicity of aPL but apparently increases their pathogenic potential. It should be kept in mind though that neither class switch nor antigen driven mutations exclude a natural *i.e.*, B1 cell origin of aPL.

2.5. Memory B-cells

Few data are available on memory B-cells in APS. Again Lieby and co-workers have provided some insight into this issue by cloning antiphospholipid specific B-cells. They showed that in non-APS patients during acute episodes of Epstein-Barr virus (EBV) infection significant numbers of aPL-producing CD27 positive B-cells were detectable which the authors regarded as memory B-cells [36]. The origin of these cells in individuals who never had any manifestations of APS remains unclear. The presence of memory B-cells capable to secrete aPL is also supported by isolated cases of the development of APS after bone-marrow transplantation from an APS donor [49]. The presence of CD27 is not restricted to memory B cells but is also found in B1 cells [14]. It should be mentioned that there is an on-going scientific debate regarding the question if a distinct CD20+CD27+CD43+ positive B cell subset represents B1 cells [50–52]. Furthermore, the ability of B1 cells to mount a T-cell independent memory response is well established [53–55]. If CD27 identifies a B1 cell subset and B1 cells can confer long-lasting immunity, the data obtained by Lieby could also be interpreted as showing an increased number of a subset of B1 cells.

3. Genetic Aspects of aPL

Genetic predisposition to the development of aPL or even APS might provide additional clues regarding the origin of these antibodies. Unfortunately, available data are scarce. There are two genome wide associations studies (GWAS) available which address the genetic associations of aPL [56,57]. Both studies explicitly do not apply to APS but focus on the presence of aPL only. While no significant association of a genetic locus with anticardiolipin antibodies was detected, several potential loci associated with antibodies against β2GPI were identified. In particular, the apolipoprotien H (*APOH*) gene itself is associated with the presence of anti-β2GPI. This confirmed previous data from candidate gene approaches which had shown that certain polymorphisms of the *APOH* locus are associated with the presence of anti-β2GPI. Of particular relevance is rs4581 causing a missense mutation Val247Leu in domain V of β2GPI [58,59]. In our hands two other missense mutations were significantly associated with the presence of anti-β2GPI. These were rs52797880 (Ile122Val) and rs8178847 (Arg135His) in domain III of β2GPI [57]. It is not known yet, if one of these polymorphisms is causally related to the development of anti-β2GPI or if they are in linkage disequilibrium to the relevant polymorphism. Another locus associated with anti-β2GPI in both GWA studies was *MACROD2*. At present, no obvious explanation for this association has been found. And finally, similar to other autoimmune diseases several possible associations of aPL and APS with the human leukocyte antigen (HLA)-locus have been reported. This has been recently reviewed in detail [60]. There are two types of studies. The first analyses the association of major histocompatibility complex (MHC) genes with APS. The second analyses the association of MHC genes with aPL. In the latter most MHC associations were found with anti-β2GPI. This again suggests that anticardiolipin and anti-β2GPI may develop along different pathways.

The available data raise an intriguing question. Apparently, there is a strong association of anti-β2GPI with the *APOH* gene and possibly a few other genes including MHC genes while no genetic association of anticardiolipin antibodies has been described, in particular not with the *APOH* locus. This obvious genetic difference implies that the origin of these two aPL species might be different. There are two potential explanations: (1) anticardiolipin and anti-β2GPI develop completely independently from each other. This appears to be unlikely considering the high coincidence of both aPL; (2) Anti-β2GPI develop preferentially in persons who have also anticardiolipin. In this case, the *APOH* polymorphisms may affect the structure of β2GPI in a way which favours autoantibody formation against the protein. The crystal structure of β2GPI revealed that the protein consists of five domains which are arranged in a J-shaped elongated form much like beads on a string [61]. Later on it was shown that β2GPI can also attain a S-shaped and a circular form [62,63]. In fact, these are the conformations that β2GPI attains when it is not bound to phospholipids. In these two conformations an epitope comprising amino acids 40–43 in domain I is hidden within the tertiary structure of the protein. Transformation to the J-shaped conformation is required in order that specific anti-β2GPI can bind to this supposedly pathogenic epitope in domain I of the protein [64]. It is conceivable that missense mutations in β2GPI affect the accessibility of this epitope to the immune system or change the overall immunogenicity of β2GPI and thereby favour the development of anti-β2GPI. This scenario requires further scientific validation.

Regarding the relationship of anticardiolipin and anti-β2GPI, we made a relevant observation in a pair of human monoclonal aPL (HL5B and HL7G) isolated from the same patient [41,43]. Both aPL have a number of identical somatic mutations, but HL7G has some additional mutations indicating that it is more advanced by antigen driven maturation. While HL5B binds to cardiolipin in the absence of cofactors and does not bind to β2GPI, HL7G in addition binds to β2GPI. This observation shows that antigen driven maturation can transform anticardiolipin specific aPL to anti-β2GPI. We do not know if this occurs regularly and can be generalized, but our data show that this is one pathway to generate anti-β2GPI. In any case, it could explain the observation that anti-β2GPI is strongly associated to the *APOH* locus, while anticardiolipin is definitely not.

4. Conclusions and Outlook

We believe that the available data in the literature very strongly support the hypothesis that aPL are natural antibodies generated by B1 cells. Figure 1 depicts the basic concept which is at present only a working model and clearly needs substantial further experimental validation. There is ample evidence that aPL with germline sequence can be pathogenic even though it is likely that antigen driven maturation can increase the pathogenic potential of aPL. In particular, the development of aPL specific for β2GPI is very probably antigen driven. There is at least one documented case that an antibody against β2GPI evolved by somatic mutation from an anticardiolipin antibody. Since it has been shown in the past that B1 cells and the antibodies produced by them can undergo antigen driven maturation, antigen driven maturation does not argue against B1 cells being a major source of aPL. If aPL derive from B1 cells it can be expected that activation via innate immune processes rather than traditional HLA-dependent pathways of adaptive immunity may play a significant role in their development.

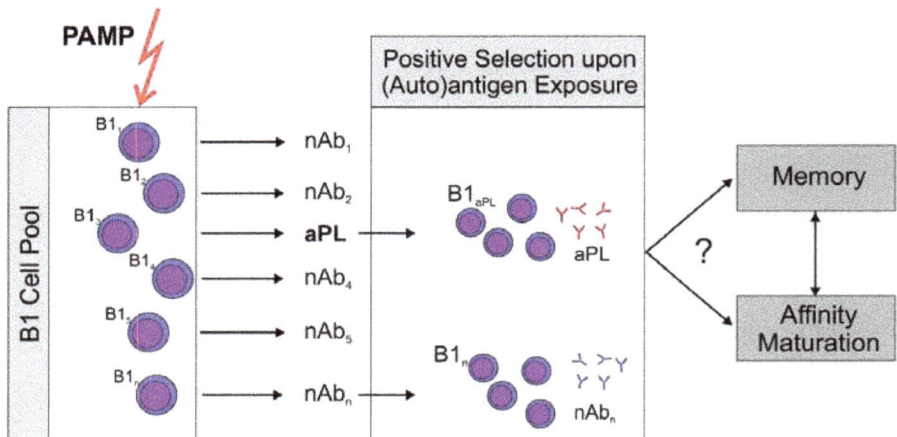

Figure 1. Proposed sequence of events leading to antiphospholipid antibodies (aPL). A non-specific stimulus by pathogen associated patterns (PAMP) which can activate pattern recognition receptors, e.g., toll-like receptors (TLR) stimulates an increase over basal antibody production by B1 cells. Subsequently, antigen producing B1 cell clones are positively selected by exposure to their (auto)antigen. This model could explain rapid aPL production induced by immunization of mice with an aPL. It should be noted that aPL themselves are able to sensitize immune cells to the action of ligands for TLR7, the receptor for single stranded RNA (ssRNA), by inducing TLR7 transcription and translocation to the endosome [21]. This could explain the role for aPL in this immunization scheme. Development of memory and antigen driven maturation have been described for B1 cells. However, there is only circumstantial evidence that this might occur with aPL producing clones.

Conflicts of Interest: The authors declare no conflict of interest.

Glossary

Germline sequence	Antibodies are encoded in the genome as every other protein. However, for certain segments (V, D, and J) of the variable chains of antibodies there are several coding gene segments. The term germline sequence refers to an antibody sequence encoded in the genome. Germline sequences can be modified by→antigen driven maturation. If an antibody has a germline encoded sequence this suggests that no antigen driven maturation has occurred, yet.

V,(D), J (or somatic) -recombination	The process of combining the gene segments for the desired V, D, and J-segments of the variable chains and of removal of surplus gene segments from the B cell genome is referred to as somatic recombination or V, (D), J-recombination. It is mediated by VDJ-recombinase, a multi-enzyme complex. Somatic recombination is the first step in antibody production that generates a huge potential diversity with more than 10^{11} theoretical combinations.
Somatic (hyper)mutation	During B-cell proliferation which occurs after antigen contact, the B-cell receptor locus can undergo an extremely high rate of somatic mutations which is several orders of magnitude greater than the spontaneous mutation rate. Most of the somatic mutations are found in specific regions of the antibody molecule, the so called hypervariable or complementarity determining regions (CDR). Somatic mutation generates B-cell clones which produce antibodies with different affinity to their antigen. The clones producing higher affinity antibodies are positively selected. Thus, somatic hypermutation is key to→antigen driven maturation of B-cell clones.
Antigen driven maturation	Also called affinity maturation, antigen driven maturation is the central process of adaptive immunity. By selecting higher affinity clones and deleting lower affinity clones, there is continuous improvement of antibody affinity to the relevant target antigen. A significant deviation of the sequence of an antibody from the known germline genes indicates antigen driven maturation.
Anti-idiotype	An idiotype describes the sum of the variable parts of a specific antibody. By this, it also includes the antigen binding site of the antibody. An anti-idiotype is an antibody that binds to a specific idiotype. In theory anti-idiotypes may mimick the antigen/epitope of the original antibody.

Abbreviations

The following non-standard abbreviations are used in this manuscript:

aPL	antiphospholipid antibody
APS	antiphospholipid syndrome
β2GPI	β2 glycoprotein I
SLE	systemic lupus erythematosus

References

1. Wassermann, A.; Neisser, A.; Bruck, C. Eine serodiagnostische Reaktion bei Syphilis. *Dtsch. Med. Wochenschr.* **1906**, *31*, 745–746. (In German) [CrossRef]
2. Landsteiner, K.; Müller, R.; Poetzl, O. Zur Frage der Komplementbindungsreaktionen bei Syphilis. *Wien. Klin. Wochenschr.* **1907**, *20*, 1565–1567. (In German)
3. Pangborn, M. Isolation and purification of a serologically active phospholipid from beef heart. *J. Biol. Chem.* **1942**, *143*, 247–256.
4. Harris, E.N.; Gharavi, A.E.; Boey, M.L.; Patel, B.M.; Mackworth-Young, C.G.; Loizou, S.; Hughes, G.R. Anticardio-lipin antibodies: Detection by radioimmunoassay and association with thrombosis in systemic lupus erythematosus. *Lancet* **1983**, *8361*, 1211–1214. [CrossRef]
5. Hughes, G.R. The anticardiolipin syndrome. *Clin. Exp. Rheumatol.* **1985**, *3*, 285–286. [PubMed]
6. Bertolaccini, M.L.; Amengual, O.; Andreoli, L.; Atsumi, T.; Chighizola, C.B.; Forastiero, R.; de Groot, P.; Lakos, G.; Lambert, M.; Meroni, P.; *et al.* 14th International Congress on Antiphospholipid Antibodies Task Force. Report on antiphospholipid syndrome laboratory diagnostics and trends. *Autoimmun. Rev.* **2014**, *13*, 917–930. [CrossRef] [PubMed]
7. Meroni, P.L.; Borghi, M.O.; Raschi, E.; Tedesco, F. Pathogenesis of the antiphospholipid syndrome: understand-ding the antibodies. *Nat. Rev. Rheumatol.* **2011**, *7*, 330–339. [CrossRef] [PubMed]
8. Ioannou, Y. The Michael Mason prize: Pathogenic antiphospholipid antibodies, stressed out antigens and the deployment of decoys. *Rheumatology* **2012**, *51*, 32–36. [CrossRef] [PubMed]
9. Poulton, K.; Rahman, A.; Giles, I. Examining how antiphospholipid antibodies activate intracellular signaling pathways: A systematic review. *Sem. Arthritis. Rheum.* **2012**, *41*, 720–736. [CrossRef] [PubMed]
10. Giannakopoulos, B.; Krilis, S.A. The pathogenesis of the antiphospholipid syndrome. *New Engl. J. Med.* **2013**, *368*, 1033–1044. [CrossRef] [PubMed]

11. Du, V.X.; Kelchtermans, H.; de Groot, P.G.; de Laat, B. From antibody to clinical phenotype, the black box of the antiphospholipid syndrome: Pathogenic mechanisms of the antiphospholipid syndrome. *Thromb. Res.* **2013**, *132*, 319–326. [CrossRef] [PubMed]

12. Merashli, M.; Noureldine, J.H.A.; Uthman, I.; Khamashta, M. Antiphospholipid syndrome: An update. *Eur. J. Clin. Investig.* **2015**, *45*, 653–662. [CrossRef] [PubMed]

13. Panda, S.; Ding, J.K. Natural antibodies bridge innate and adaptive immunity. *J. Immunol.* **2015**, *194*, 13–20. [CrossRef] [PubMed]

14. Rothstein, T.L.; Griffin, D.O.; Holodick, N.E.; Quach, T.D.; Kaku, H. Human B-1 cells take the stage. *Ann. New York Acad. Sci.* **2013**, *1285*, 97–114. [CrossRef] [PubMed]

15. Youinou, P.; Renaudineau, Y. The antiphospholipid syndrome as a model for B cell-induced autoimmune diseases. *Thromb. Res.* **2004**, *114*, 363–369. [CrossRef] [PubMed]

16. Merrill, J.T. Do antiphospholipid antibodies develop for a purpose? *Curr. Rheumatol. Rep.* **2006**, *8*, 109–113. [CrossRef] [PubMed]

17. Bakimer, R.; Fishman, P.; Blank, M.; Sredni, B.; Djaldetti, M.; Shoenfeld, Y. Induction of primary antiphospholipid syndrome in mice by immunization with a human monoclonal anticardiolipin antibody (H-3). *J. Clin. Investig.* **1992**, *89*, 1558–1563. [CrossRef] [PubMed]

18. Shoenfeld, Y. Idiotypic induction of autoimmunity: A new aspect of the idiotypic network. *FASEB J.* **1994**, *8*, 1296–1301. [PubMed]

19. Pierangeli, S.S.; Harris, E.N. Induction of phospholipid-binding antibodies in mice and rabbits by immunization with human β2 glycoprotein 1 or anticardiolipin antibodies alone. *Clin. Exp. Immunol.* **1993**, *93*, 269–272. [CrossRef] [PubMed]

20. Pierangeli, S.S.; Liu, S.W.; Anderson, G.; Barker, J.H.; Harris, E.N. Thrombogenic properties of murine anti-cardiolipin antibodies induced by beta 2 glycoprotein 1 and human immunoglobulin G antiphospholipid antibodies. *Circulation* **1996**, *94*, 1746–1751. [CrossRef] [PubMed]

21. Prinz, N.; Clemens, N.; Strand, D.; Pütz, I.; Lorenz, M.; Daiber, A.; Stein, P.; Degreif, A.; Radsak, M.; Schild, H.; *et al.* Antiphospholipid antibodies induce translocation of TLR7 and TLR8 to the endosome in human monocytes and plasmacytoid dendritic cells. *Blood* **2011**, *118*, 2322–2332. [CrossRef] [PubMed]

22. Fukui, R.; Saitoh, S.; Kanno, A.; Onji, M.; Shibata, T.; Ito, A.; Onji, M.; Matsumoto, M.; Akira, S.; Yoshida, N.; *et al.* Unc93B1 restricts systemic lethal inflammation by orchestrating Toll-like receptor 7 and 9 trafficking. *Immunity* **2011**, *35*, 69–81. [CrossRef] [PubMed]

23. Yokogawa, M.; Takaishi, M.; Nakajima, K.; Kamijima, R.; Fujimoto, C.; Kataoka, S.; Terada, Y.; Sano, S. Epicutaneous application of toll-like receptor 7 agonists leads to systemic autoimmunity in wild-type mice: A new model of systemic Lupus erythematosus. *Arthritis Rheumatol.* **2014**, *66*, 694–706. [CrossRef] [PubMed]

24. Takagi, H.; Arimura, K.; Uto, T.; Fukaya, T.; Nakamura, T.; Choijookhuu, N.; Hishikawa, Y.; Sato, K. Plasmacytoid dendritic cells orchestrate TLR7-mediated innate and adaptive immunity for the initiation of autoimmune inflammation. *Sci. Rep.* **2016**, *6*. [CrossRef] [PubMed]

25. Gotoh, M.; Matsuda, J. Induction of anticardiolipin antibody and/or lupus anticoagulant in rabbits by immunization with lipoteichoic acid, lipopolysaccharide and lipid, A. *Lupus* **1996**, *5*, 593–597. [CrossRef] [PubMed]

26. Abdel-Wahab, N.; Lopez-Olivo, M.A.; Pinto-Patarroyo, G.P.; Suarez-Almazor, M.E. Systematic review of case reports of antiphospholipid syndrome following infection. *Lupus* **2016**. in press. [CrossRef]

27. Gharavi, A.E.; Pierangeli, S.S.; Espinola, R.G.; Liu, X.; Colden-Stanfield, M.; Harris, E.N. Antiphospholipid antibodies induced in mice by immunization with a cytomegalovirus-derived peptide cause thrombosis and activation of endothelial cells *in vivo*. *Arthritis Rheum.* **2002**, *46*, 545–552. [CrossRef] [PubMed]

28. Blank, M.; Krause, I.; Fridkin, M.; Keller, N.; Kopolovic, J.; Goldberg, I.; Tobar, A.; Shoenfeld, Y. Bacterial induction of autoantibodies to beta2-glycoprotein-I accounts for the infectious etiology of antiphospholipid syndrome. *J. Clin. Investig.* **2002**, *109*, 797–804. [CrossRef] [PubMed]

29. Gharavi, A.E.; Pierangeli, S.S.; Harris, E.N. Viral origin of antiphospholipid antibodies: Endothelial cell activa-tion and thrombus enhancement by CMV peptide-induced APL antibodies. *Immunobiology* **2003**, *207*, 37–42. [CrossRef] [PubMed]

30. Shoenfeld, Y.; Blank, M.; Cervera, R.; Font, J.; Raschi, E.; Meroni, P.L. Infectious origin of the antiphospholipid syndrome. *Ann. Rheum. Dis.* **2006**, *65*, 2–6. [CrossRef] [PubMed]

31. Martin, E.; Winn, R.; Nugent, K. Catastrophic antiphospholipid syndrome in a community-acquired methicillin-resistant Staphylococcus aureus infection: A review of pathogenesis with a case for molecular mimicry. *Autoimmun. Rev.* **2011**, *10*, 181–188. [CrossRef] [PubMed]
32. Justo, D.; Finn, T.; Atzmony, L.; Guy, N.; Steinvil, A. Thrombosis associated with acute cytomegalo-virus infection: A meta-analysis. *Eur. J. Intern. Med.* **2011**, *22*, 195–199. [CrossRef] [PubMed]
33. Uthman, I.; Tabbarah, Z.; Gharavi, A.E. Hughes syndrome associated with cytomegalovirus infection. *Lupus* **1999**, *8*, 775–777. [CrossRef] [PubMed]
34. Nakayama, T.; Akahoshi, M.; Irino, K.; Kimoto, Y.; Arinobu, Y.; Niiro, H.; Tsukamoto, H.; Horiuchi, T.; Akashi, K. Transient antiphospholipid syndrome associated with primary cytomegalovirus infection: A case report and literature review. *Case Rep. Rheumatol.* **2014**, *2014*. [CrossRef] [PubMed]
35. Giles, I.P.; Haley, J.D.; Nagl, S.; Isenberg, D.A.; Latchman, D.S.; Rahman, A. A systematic analysis of sequences of human antiphospholipid and anti-b2-glycoprotein I antibodies: The importance of somatic mutations and certain sequence motifs. *Semin. Arthritis Rheum.* **2003**, *32*, 246–265. [CrossRef] [PubMed]
36. Elkon, K.; Casali, P. Nature and functions of autoantibodies. *Nat. Clin. Pract. Rheumatol.* **2008**, *4*, 491–498. [CrossRef] [PubMed]
37. Lieby, P.; Soley, A.; Levallois, H.; Hugel, B.; Freyssinet, J.-M.; Cerutti, M.; Pasquali, J.-L.; Martin, T. The clonal analysis of anticardiolipin antibodies in a single patient with primary antiphospholipid syndrome reveals extreme antibody heterogeneity. *Blood* **2001**, *97*, 3820–3828. [CrossRef] [PubMed]
38. Lieby, P.; Soley, A.; Knapp, A.-M.; Cerutti, M.; Freyssinet, J.-M.; Pasqualit, J.-L.; Martin, T. Memory B cells producing somatically mutated antiphospholipid antibodies are present in healthy individuals. *Blood* **2003**, *102*, 2459–2465. [CrossRef] [PubMed]
39. Pasquali, J.-L.; Nehme, H.; Korganow, A.-S.; Martin, T. Antiphospholipid antibodies: Recent progresses on their origin and pathogenicity. *Joint Bone Spine* **2004**, *71*, 172–174. [CrossRef] [PubMed]
40. Lieby, P.; Poindron, V.; Roussi, S.; Klein, C.; Knapp, A.M.; Garaud, J.C.; Cerutti, M.; Martin, T.; Pasquali, J.L. Patho-genic antiphospholipid antibody: An antigen-selected needle in a haystack. *Blood* **2004**, *104*, 1711–1715. [CrossRef] [PubMed]
41. von Landenberg, C.; Lackner, K.J.; von Landenberg, P.; Lang, B.; Schmitz, G. Isolation and character-rization of two human monoclonal antiphospholipid IgG from patients with autoimmune disease. *J. Autoimmun.* **1999**, *13*, 215–223. [CrossRef] [PubMed]
42. Buschmann, C.; Fischer, C.; Ochsenhirt, V.; Neukirch, C.; Lackner, K.J.; von Landenberg, P. Generation and characterization of three monoclonal IgM antiphospholipid antibodies recognizing different phospholipid antigens. *Ann. N. Y. Acad. Sci.* **2005**, *1051*, 240–254. [CrossRef] [PubMed]
43. Prinz, N.; Häuser, F.; Lorenz, M.; Lackner, K.J.; von Landenberg, P. Structural and functional characterization of a human IgG monoclonal antiphospholipid antibody. *Immunobiology* **2011**, *216*, 145–151. [CrossRef] [PubMed]
44. Prinz, N.; Clemens, N.; Canisius, A.; Lackner, K.J. Endosomal NADPH-oxidase is critical for induction of the tissue factor gene in monocytes and endothelial cells. Lessons from the antiphospholipid syndrome. *Thromb. Haemost.* **2013**, *109*, 525–531. [CrossRef] [PubMed]
45. Müller-Calleja, N.; Köhler, A.; Siebald, B.; Canisius, A.; Orning, C.; Radsak, M.; Stein, P.; Mönnikes, R.; Lackner, K.J. Cofactor-independent antiphospholipid antibodies activate the NLRP3-inflammasome via endosomal NADPH-oxidase: Implications for the antiphospholipid syndrome. *Thromb. Haemost.* **2015**, *113*, 1071–1083. [CrossRef] [PubMed]
46. Manukyan, D.; Müller-Calleja, N.; Jäckel, S.; Luchmann, K.; Mönnikes, R.; Kiouptsi, K.; Reinhardt, C.; Jurk, K.; Walter, U.; Lackner, K.J. Cofactor Independent Human Antiphospholipid Antibodies Induce Venous Thrombosis in Mice. *J. Thromb. Haemost.* **2016**, *14*, 1011–1020. [PubMed]
47. Girardi, G.; Berman, J.; Redecha, P.; Spruce, L.; Thurman, J.M.; Kraus, D.; Hollmann, T.J.; Casali, P.; Caroll, M.C.; Wetsel, R.A.; *et al.* Complement C5a receptors and neutrophils mediate fetal injury in the antiphospholipid syndrome. *J. Clin. Investig.* **2003**, *112*, 1644–1654. [CrossRef] [PubMed]
48. Ikematsu, W.; Luan, F.L.; La Rosa, L.; Beltrami, B.; Nicoletti, F.; Buyon, J.P.; Meroni, P.L.; Balestrieri, G.; Casali, P. Human anticardiolipin monoclonal autoantibodies cause placental necrosis and fetal loss in BALB/c mice. *Arthritis Rheum.* **1998**, *41*, 1026–1039.

49. Ritchie, D.S.; Sainani, A.; D'Souza, A.; Grigg, A.P. Passive donor-to-recipient transfer of antiphospholipid syndrome following allogeneic stem-cell transplantation. *Am. J. Hematol.* **2005**, *79*, 299–302. [CrossRef] [PubMed]

50. Griffin, D.O.; Holodick, N.E.; Rothstein, T.L. Human B1 cells in umbilical cord and adult peripheral blood express the novel phenotype CD20+CD27+CD43+CD70−. *J. Exp. Med.* **2011**, *208*, 67–80. [CrossRef] [PubMed]

51. Tangye, S.G. To B1 or not to B1: That really is still the question! *Blood* **2013**, *121*, 5109–5110. [CrossRef] [PubMed]

52. Inui, M.; Hirota, S.; Hirano, K.; Fujii, H.; Sugahara-Tobinai, A.; Ishii, T.; Harigae, H.; Takai, T. Human CD43+ B cells are closely related not only to memory B cells phenotypically but also to plasmablasts developmentally in healthy individuals. *Int. Immunol.* **2015**, *27*, 345–355. [CrossRef] [PubMed]

53. Allugupalli, K.R.; Leong, J.M.; Woodland, R.T.; Muramatsu, M.; Honjo, T.; Gerstein, R.M. B1b lymphocytes confer T cell-independent long-lasting immunity. *Immunity* **2004**, *21*, 379–390. [CrossRef] [PubMed]

54. Yang, Y.; Ghosn, E.E.; Cole, L.E.; Obukhanych, T.V.; Sadate-Ngatchou, P.; Vogel, S.N.; Herzenberg, L.A.; Herzenberg, L.A. Antigen-specific memory in B-1a and its relationship to natural immunity. *Proc. Natl. Acad. Sci. USA* **2012**, *109*, 5388–5393. [CrossRef] [PubMed]

55. Yang, Y.; Ghosn, E.E.; Cole, L.E.; Obukhanych, T.V.; Sadate-Ngatchou, P.; Vogel, S.N.; Herzenberg, L.A.; Herzenberg, L.A. Antigen-specific antibody responses in B-1a and their relationship to natural immunity. *Proc. Natl. Acad. Sci. USA* **2012**, *109*, 5382–5387. [CrossRef] [PubMed]

56. Kamboh, M.I.; Wang, X.; Kao, A.H.; Barmada, M.M.; Clarke, A.; Ramsey-Goldman, R.; Manzi, S.; Demirci, F.Y. Genome-wide association study of antiphospholipid antibodies. *Autoimmun. Dis.* **2013**, *2013*. [CrossRef] [PubMed]

57. Müller-Calleja, N.; Rossmann, H.; Müller, C.; Wild, P.; Blankenberg, S.; Pfeiffer, N.; Binder, H.; Beutel, M.E.; Manukyan, D.; Zeller, T.; *et al.* Antiphospholipid antibodies in a large population-based cohort: genome-wide associations and effects on monocyte gene expression. *Thromb. Haemost.* **2016**, *116*. [CrossRef] [PubMed]

58. Hirose, N.; Williams, R.; Alberts, A.R.; Furie, R.A.; Chartash, E.K.; Jain, R.I.; Sison, C.; Lahita, R.G.; Merrill, J.T.; Cucurull, E.; *et al.* A role for the polymorphism at position 247 of the beta2-glyco-protein I gene in the generation of anti-beta2-glycoprotein I antibodies in the antiphospholipid syndrome. *Arthritis Rheum.* **1999**, *42*, 1655–1661. [CrossRef]

59. Chamorro, A.J.; Marcos, M.; Mirón-Canelo, J.A.; Cervera, R.; Eapinosa, G. Val247Leu β2-glycoprotein-I allelic variant is associated with antiphospholipid syndrome: Systematic review and meta-analysis. *Autoimmun. Rev.* **2012**, *11*, 705–712. [CrossRef] [PubMed]

60. Sebastiani, G.D.; Iuliano, A.; Cantarini, L.; Galeazzi, M. Genetic aspects of the antiphospholipid syndrome: An update. *Autoimmun. Rev.* **2016**, *15*, 433–439. [CrossRef] [PubMed]

61. Schwarzenbacher, R.; Zeth, K.; Diederichs, K.; Gries, A.; Kostner, G.M.; Laggner, P.; Prassl, R. Crystal structure of human beta2-glycoprotein I: Implications for phospholipid binding and the antiphospholipid syndrome. *EMBO J.* **1999**, *18*, 6228–6239. [CrossRef] [PubMed]

62. Hammel, M.; Kriechbaum, M.; Gries, A.; Kostner, G.M.; Laggner, P.; Prassl, R. Solution structure of human and bovine beta(2)-glycoprotein I revealed by small-angle X-ray scattering. *J. Mol. Biol.* **2002**, *321*, 85–97. [CrossRef]

63. Agar, C.; van Os, G.M.; Mörgelin, M.; Sprenger, R.R.; Marquart, J.A.; Urbanus, R.T.; Derksen, R.H.; Meijers, J.C.; de Groot, P.G. Beta2-glycoprotein I can exist in 2 conformations: Implications for our understanding of the antiphospholipid syndrome. *Blood* **2010**, *116*, 1336–1343. [CrossRef] [PubMed]

64. Ninivaggi, M.; Kelchtermans, H.; Lindhout, T.; de Laat, B. Conformation of beta2glycoprotein I and its effect on coagulation. *Thromb. Res.* **2012**, *130*, S33–S36. [CrossRef] [PubMed]

antibodies

MDPI

Review

The Significance of Anti-Beta-2-Glycoprotein I Antibodies in Antiphospholipid Syndrome

Anna Brusch

Department of Clinical Immunology, PathWest, Sir Charles Gairdner Hospital, Perth WA 6009, Australia; anna.brusch@health.wa.gov.au; Tel.: +61-8-9346-3333

Academic Editor: Ricard Cervera
Received: 26 April 2016; Accepted: 3 June 2016; Published: 8 June 2016

Abstract: Antiphospholipid syndrome (APS) is a thrombophilic disorder that classically presents with vascular thrombosis and/or obstetric complications. APS is associated with antiphospholipid antibodies: a heterogeneous group of autoantibodies that are directed against membrane phospholipids in complex with phospholipid-binding proteins. Beta-2-glycoprotein I (B2GPI) binds anionic phospholipids and is considered to be the predominant antigen in APS and antibodies against B2GPI (anti-B2GPI) are recognised in the laboratory criteria for APS diagnosis. This review focuses on the part played by anti-B2GPI in the pathogenesis of APS, their associations with different clinical phenotypes of the disorder and new avenues for refining the diagnostic potential of anti-B2GPI testing.

Keywords: antiphospholipid syndrome; antiphospholipid antibody; beta-2 glycoprotein I; isotypes; domain specific antibodies

---

1. Introduction

Antiphospholipid syndrome (APS) is an autoimmune thrombophilia that can occur in isolation or as part of an underlying systemic autoimmune disorder, such as systemic lupus erythematosus (SLE) [1]. It has a wide range of potential clinical manifestations: most commonly presenting with arterial or venous thrombosis, which may occur in the absence of other risk factors. Recurrent thrombosis can occur in some patients. Obstetric complications such as recurrent early miscarriages, preeclampsia and late fetal loss also form a subset of the disorder, known as obstetric APS [2]. While these clinical phenotypes are typical of APS, there are no pathognomic features that can secure the diagnosis of APS on clinical grounds alone. Rather, the diagnosis is made by the combination of clinical features together with supportive laboratory findings. This relies on the accurate identification and measurement of antiphospholipid antibodies (aPL). A number of aPL have been described in APS, however, only three are included in the current consensus guidelines regarding diagnosis [3]. These are lupus anticoagulant (LA), anticardiolipin antibodies (aCL) and anti-beta-2 glycoprotein I antibodies (anti-B2GPI). This review will focus on anti-B2PI and their role in APS in terms of their relationship to the putative pathogenesis of the disorder and their clinical associations. There are also areas of ongoing doubt such as the relative significance of particular anti-B2GPI isotypes and new areas of investigation including the potential for domain specific antibodies to refine the diagnostic value of anti-B2GPI testing.

2. Anti-B2GPI: Antibodies against an Enigmatic, Multi-Purpose Target

A role for anti-B2GPI in the pathogenesis of APS has been demonstrated in *in vivo* animal models [4]. It is hypothesised that anti-B2GPI bind to membrane-bound B2GPI complexed with anionic phospholipids expressed on the surface of a range of cells involved in the coagulation cascade which triggers cellular signaling events culminating in procoagulant effects such as modification of endothelial cells, potentiation of platelet aggregation and interference with plasma clotting components [5]. However, despite B2GPI being the predominant target in APS pathogenesis, its precise physiological function remains elusive.

The normal function of B2GPI has largely been inferred from scrutinising its complex protein structure. B2GPI contains five domains composed of repeating stretches of about 60 amino acids, similar to other proteins of the complement control protein superfamily [6]. The 5th domain contains a C-terminal extension and an additional disulphide bond that confers a positive charge resulting in an affinity for anionic phospholipids. The crystal structure of B2GPI was first elucidated in the late 1990s and demonstrated a stretched arrangement of domains 1–4, with the 5th domain protruding at a right angle giving an appearance resembling the letter 'J' or a hockey stick. Subsequent analysis by small angle X-ray scattering experiments suggested that in solution, B2GPI adopted an 'S-shaped' conformation [7]. More recently, electron microscopy studies also indicate that the structure of B2GPI is not limited to a single conformation. Rather, B2GPI can assume a different geometry in fluid phase which may alter its potential to interact with autoantibodies [8]. By electron microscopy analysis, B2GPI was found to assume a circular conformation in plasma with domains 1 and 5 opposed. In this form, the site(s) for autoantibody binding are shielded. Binding of anti-B2GPI to membrane-bound B2GPI stabilises its J-shaped structure and augments B2GPI's interaction with membrane phospholipids which is hypothesised to potentiate B2GPI's signaling through other transmembrane and intracellular ligands. These include toll-like receptors; TLR2 and TLR4, annexin A2 and LRP8 [6]. Signaling via these molecules mediates prothrombotic cellular actions.

In patients with APS, thrombotic events occur with increasing frequency in the presence of other prothrombotic risk factors such as infection. How these multiple 'hits' align to result in thrombosis is likely to be complex and multifactorial. However, recent studies have started to shed light on this area by indicating a potential interplay between B2GPI and various elements of the immune system during infection. For example, the positively charged sites in domain 5 of B2GPI confer an affinity for negatively charged cell membranes and are also thought to result in interactions with bacteria that might trigger innate immune responses. Indeed, peptides derived from domain 5 have been shown to display potent antibacterial activity against a variety of bacteria [9]. Other studies have shown that B2GPI interacts directly with lipopolysaccharide resulting in a complex that can be recognised and internalised by macrophages [10]. In this way, B2GPI's interactions with numerous ligands give it the capacity to both sense and respond to signals and provides a potential 'meeting point' for various thrombophilic stimuli to converge.

B2GPI may also play a regulatory role in important immune pathways that could in turn be disrupted by the presence of anti-B2GPI. For example, a recent study demonstrated that the elongated, membrane-bound form of B2GPI acts as a binding site for the complement protein, C3 [11]. The complex of B2GPI and C3 may then in turn serve a dual purpose. In addition to opsonising apoptotic cells, C3 binding to B2GPI provides a binding site for factor H which then mediates degradation of C3 via the activity of factor I. There are suggestions that complement dysregulation plays a part in APS as evidenced by data from mouse models of APS showing that inhibition of C3 activity can prevent fetal loss [12,13]. Furthermore, inhibition of complement component C5 by eculizumab has been used to treat patients with catastrophic APS [14]. However, how the complement pathway is affected by the presence of anti-B2GPI is yet to be determined.

3. Testing for Anti-B2GPI: An Evolving Component of the Laboratory Criteria for APS

The identification of B2GPI as a target for the pathogenic pathways of APS has prompted studies to look at the utility of anti-B2GPI in APS diagnosis. Consensus guidelines for APS were originally compiled in the 1990s and pertained to lupus anticoagulant (LA) and anticardiolipin antibodies (aCL) [15]. However, they were subsequently updated in 2006 on account of several studies indicating a role for anti-B2GPI to identify APS patients with both vascular and obstetric APS [3]. Estimates of the prevalence of anti-B2GPI in APS vary and this may be attributable to the heterogeneity of patient populations as well as differences in assays used. Several studies have reported isolated anti-B2GPI and this may account for between 11% and 27% of APS patients [16,17]. Isolated anti-B2GPI seems to occur more frequently than either isolated LA or isolated aCL; accounting for 75% of APS patients who had a single antiphospholipid antibody in one study [17]. Analysis of isolated anti-B2GPI has also been performed in the specific context of obstetric APS. Among 500 healthy women who were prospectively screened for aPL in early pregnancy, 4% were found to have anti-B2GPI without other aPL. Pre-eclampsia and eclampsia occurred significantly more frequently among these women compared with aPL-negative women raising a potentially important association with this particular obstetric complication [18]. A recent systematic review and meta-analysis of the risk of thrombotic events according to aPL type indicated that the association of anti-B2GPI is greater for arterial events than for venous thrombosis in patients without systemic lupus erythematosus (SLE) [19].

4. Significance of Anti-B2GPI Isotypes: An Ongoing Area of Contention

Following recognition of the importance of anti-B2GPI in the development of APS, further study has been directed towards understanding the significance of particular isotypes of anti-B2GPI. This issue was addressed at the International Congress on Antiphospholipid Antibodies in 2010 and an update to the international consensus guidelines on anticardiolipin and anti-B2GPI testing was published in 2012 [20]. These guidelines conclude that the evidence for an association between anti-B2GPI and APS is strongest for the IgG isotype. The guidelines acknowledge that data continues to build for IgA anti-B2GPI, but currently testing for this isotype is recommended in patients with negative IgG and IgM anti-B2GPI and in whom APS is still suspected. However, debate continues as new studies emerge that discuss the relative merits of the three anti-B2GPI isotypes.

4.1. Utility of IgM Anti-B2GPI

The value of IgM anti-B2GPI testing in APS has been supported by reports of patients with isolated IgM anti-B2GPI as the sole aPL. For example, in an Italian study of 64 primary APS patients (diagnosed on clinical grounds) with only one aPL detected, over 50% were positive for IgM anti-B2GPI. Approximately two-thirds of those with isolated IgM anti-B2GPI had obstetric manifestations of APS, perhaps indicating a possible association of IgM anti-B2GPI and this particular subtype of APS [17].

By contrast, several other studies have reported a stronger association between IgG anti-B2GPI and clinical manifestations of APS compared with IgM anti-B2GPI which has cast considerable doubt on the role of IgM anti-B2GPI in the first-line assessment of APS. In one of the first studies to examine this issue, Lakos *et al.* reported no association between IgM anti-B2GPI and typical APS manifestations, including venous thrombosis and miscarriage, in contrast to IgG anti-B2GPI [21]. Another study looked at the ability of anti-B2GPI isotypes to stratify thrombotic risk in patients with lupus anticoagulant, and found that those with IgG anti-B2GPI had a higher incidence of thrombosis whereas there was no association for IgM anti-B2GPI [22]. A recent study from Western Australia assessed the clinical phenotype of 128 hospital patients who had tested positive for at least one anti-B2GPI isotype [23]. There was a higher proportion of patients with a history of unprovoked venous and arterial thrombosis among those who were IgG anti-B2GPI positive compared to those who lacked IgG anti-B2GPI. This was not the case for IgM or IgA anti-B2GPI. Similarly, median IgG anti-B2GPI levels were higher among patients with unprovoked thrombosis compared to those with clinical events less in keeping with APS. The reverse trend was observed for IgM anti-B2GPI antibody levels suggesting that IgM anti-B2GPI had the least robust clinical association with potential APS.

Indeed, some data suggests that IgM anti-B2GPI may actually play a protective role against disease in some situations. In a study of nearly 800 patients with SLE, those with IgM anti-B2GPI had a lower incidence of lupus nephritis, hypertension and renal impairment. Furthermore, IgM anti-B2GPI did not associate with arterial or venous thrombosis in contrast to IgG and IgA anti-B2GPI isotypes [24].

Given mounting evidence that IgM anti-B2GPI may not be strongly associated with APS, there is concern that its continued inclusion as a first line aPL test for APS is no longer warranted. Indeed the data currently suggest that IgM anti-B2GPI is less likely than other isotypes to associate with APS. However, a persistently positive IgM anti-B2GPI result could help substantiate a diagnosis of APS for individual patients presenting with thrombotic or obstetric complications. Therefore, further discussion regarding how to balance these conflicting ideas from the standpoint of the diagnostic laboratory is required to achieve a pragmatic testing strategy that is helpful to clinicians.

4.2. Associations of IgA Anti-B2GPI

Of the three anti-B2GPI isotypes, the role for IgA anti-B2GPI testing remains the most contentious and ambiguous. The first reports of IgA anti-B2GPI's potential role in APS diagnosis emerged in the late 1990s from case reports of patients with APS and IgA anti-B2GPI with an otherwise negative aPL profile and also from relatively small patient cohort studies demonstrating an association between IgA anti-B2GPI and APS manifestations [21,25–27]. Since then, numerous other studies examining this area have been published, yet uncertainty remains regarding key questions such as the prevalence of IgA anti-B2GPI in health and disease. After nearly 20 years of study, the prevalence of IgA anti-B2GPI in primary and secondary APS has not yet been definitively established, with a variety of reported estimates ranging from 14% to 72% [28]. Methodological differences between studies such as differences in patient demographics and clinical phenotypes (e.g., patient ethnicities, primary *vs.* secondary APS, thrombotic *vs.* obstetric manifestations, *etc.*), varying assay methods and small patient numbers have added complexity to the interpretation of the data.

After initial reports of a positive association between IgA anti-B2GPI and APS manifestations, several studies published data disputing a role for this isotype in APS diagnosis. Danowski *et al.* reported that among a group of approximately 400 SLE patients with APS, IgA anti-B2GPI did not associate with any manifestation of APS in contrast to IgG and IgM anti-B2GPI [16]. Samarkos *et al.* reported similar findings among patients with primary APS and SLE [29].

Given the conflicting data regarding the utility of IgA anti-B2GPI testing in APS, consensus guidelines have not definitively included or excluded the isotype from testing algorithms. Rather, it is suggested that testing remains an option for individual patients who are negative for other aPL. This standpoint may be debated again when the guidelines are next reviewed in light of recent studies continuing to support IgA anti-B2GPI's role in APS including from larger patient cohort studies.

Indeed, one of the largest studies to date examining the utility of IgA anti-B2GPI testing included nearly 6000 patients [30]. The majority of this group (5098 patients) were patients being assessed for potential APS at the Antiphospholipid Standardisation Laboratory in Texas. The remainder (approximately 900 patients) were drawn from two established cohorts of patients with SLE. The overall prevalence of IgA in the non-SLE group was low at <1% compared with approximately 20% in the SLE patient groups. Isolated IgA anti-B2GPI prevalence was <0.5% and 5% in the non-SLE and SLE groups respectively. A considerable number of patients with IgA anti-B2GPI had at least one APS-related clinical manifestation which included classical APS presentations as well as non-classical features such as thrombocytopenia and livedo reticularis. An interesting finding to emerge among the groups of patients with SLE was that IgA anti-B2GPI was significantly associated with arterial thrombosis but not venous thrombosis. A recent study from Spain also reported a stronger association with arterial thrombosis for IgA anti-B2GPI compared with IgG or IgM anti-B2GPI among 156 patients who met clinical criteria for APS [31]. This study also reported significantly different prevalence for IgA anti-B2GPI among patients with primary APS compared with APS associated with systemic autoimmune disease. These findings could conceivably indicate a variable role for IgA anti-B2GPI in different pathogenic pathways of APS resulting in distinct clinical phenotypes.

A major challenge to studying aPL and their associations with APS relates to the variability in diagnostic assays and lack of assay standardisation. This is particularly relevant to IgA anti-B2GPI which exhibits greater variability in results between assays compared to IgG and IgM isotypes.

For example, a recent study compared results for IgA anti-B2GPI from approximately 70 patients using seven different commercial enzyme-linked immunosorbent assay (ELISA) kits and demonstrated substantial differences in sensitivity and specificity between assays [32]. Similar findings were also discussed in the Antiphospholipid Antibodies Task Force report on laboratory diagnostics pertaining to data from cohorts of APS and SLE patients. Comparisons of assays for each anti-B2GPI isotype identified a lower level of agreement between assays for IgA- compared with IgG- and IgM anti-B2GPI [33].

5. Anti-B2GPI Outside of APS: Questions Regarding Specificity

The precise role of anti-B2GPI isotypes in APS remains incompletely resolved and this can often lead to clinical uncertainty when interpreting the significance of a positive anti-B2GPI result. As discussed already, the prevalence of anti-B2GPI antibodies among patients with different diseases has not been firmly established but would provide important data when considering the specificity of anti-B2GPI antibodies. It is well known that anticardiolipin antibodies can be seen in conditions other than APS. Medications, infections and other illnesses have been reported in association with aCL which are often transient and unsustained [34]. Less is known about the associations and significance of anti-B2GPI outside of APS but several studies have addressed this issue in selected patient populations with diseases other than APS. For example, infections are thought to be an important trigger of aPL production and this phenomenon is thought to be due in part to molecular mimicry [34]. Indeed, a study of patients in South Africa found anti-B2GPI antibodies in 6-8% of patients of HIV, syphilis and malaria and in 89% and 30% respectively of patients with leprosy and hepatitis C [35].

With regard to disease associations of particular isotypes, an increased prevalence of IgA anti-B2GPI has been reported in a variety of disorders such as autoimmune hepatitis, coeliac disease, metabolic syndrome, and haemodialysed patients with end stage renal failure [36–39]. The significance of these associations remains unclear as the presence of anti-B2GPI is not always associated with APS manifestations. However, in some circumstances, the presence of IgA anti-B2GPI may confer a worse prognosis of the underlying disease. For example, in the case of end stage renal failure patients receiving haemodialysis, IgA anti-B2GPI were an independent risk factor for mortality and antibody levels fell in patients who received a renal transplant [39,40].

6. Obstetric APS and Anti-B2GPI

Obstetric APS represents a subset of APS and is thought to be mediated by distinct pathogenic mechanisms. As well as thrombosis of placental vessels, non-thrombotic mechanisms are also thought to be important. These include binding of anti-B2GPI to trophoblasts resulting in modulation of trophoblast proliferation and growth. Anti-B2GPI may also affect endometrial cells in the decidua that might impede implantation. Complement activation and enhanced apoptosis of embryonic and placental cells may also play a role in the pathways that lead to recurrent early miscarriage, fetal loss, pre-eclampsia and placental insufficiency [5]. Consequently, it has been hypothesised that patients with obstetric APS may differ in their aPL profiles compared with patients with predominantly vascular APS. As with vascular APS, it is difficult to combine evidence from obstetric APS studies on account of variability in study design, clinical case definition, range of aPL tested variance of laboratory reference ranges from consensus guideline recommendations and other constraints. This has been discussed in detail by the Antiphospholipid Antibody Task Force who reported their findings from the 14th International Congress on antiphospholipid antibodies [41].

The available data continue to show conflicting results regarding the utility of anti-B2GPI in obstetric APS. A meta-analysis of studies pertaining to placental problems in pregnancy (comprising late fetal loss, preeclampsia, placental abruption or intrauterine growth restriction), identified two cohort studies which demonstrated an association between anti-B2GPI and preeclampsia and late fetal loss [18,42,43]. One cohort study demonstrated an association with intrauterine growth restriction. However, there were four other case-control studies which did not confirm these associations [44–47].

The meta-analysis concluded that there was insufficient data to establish a significant link between anti-B2GPI and pregnancy morbidity. By contrast, LA was found to associate with placenta mediated complications with an odds ratio of approximately 10 for late fetal loss among cohort studies. More recently, a large multicentre prospective study including nearly 600 cases of fetal death after 20 weeks gestation found that elevated levels of IgM and IgG anti-B2GPI were associated with an increased risk of stillbirth, with odds ratios of 2 and 3 respectively [48]. One factor that has hampered direct comparison of various aPL in obstetric APS has been the heterogeneity of antibodies tested among different studies. A limited number of studies have included triple testing of LA, aCL and anti-B2GPI and indicate that positivity for all three antibodies is a risk factor for pregnancy morbidity [41,49].

The majority of studies in obstetric APS have included testing for IgG and/or IgM anti-B2GPI. However, there have been several studies that have examined the associations of IgA anti-B2GPI in relation to obstetric outcomes. Perhaps unsurprisingly, the data shows conflicting results. For example, a retrospective study in the early 2000s found that pregnant women with APS, preeclampsia and autoimmune disease had significantly higher IgA anti-B2GPI levels compared with a group of pregnant patients with diabetes and women with normal pregnancies [50]. Another study also found higher IgA anti-B2GPI levels among women with recurrent spontaneous miscarriages and fetal loss [51]. However, a study looking at anti-B2GPI antibodies in 84 women with primary APS, unexplained pregnancy morbidity and SLE found a low prevalence of IgA anti-B2GPI and no significant clinical associations [52].

Overall, the evidence relating to the significance of anti-B2GPI in obstetric APS is hindered by differences in the clinical and epidemiological characteristics of patients studied, as well as methodological differences in assays used. Longitudinal, prospective studies are needed in this field to help resolve the ambiguities regarding the prevalence and significance of anti-B2GPI in obstetric APS.

7. Future Directions for Anti-B2GPI Testing

7.1. Refining the Target of Anti-B2GPI Testing: The Role of Domain Specific Antibodies

There are many difficulties in studying anti-B2GPI antibodies and one of the key remaining questions is how to identify patients with the highest risk of thrombosis or obstetric morbidity—*i.e.*, why do some patients with anti-B2GPI develop APS and others do not? A current area of assay development aims to distinguish between different subtypes of anti-B2GPI antibody that may differ in their pathogenicity. Standard anti-B2GPI assays do not differentiate autoantibodies against different domains of B2GPI. There is emerging evidence to suggest that determining the target domain of anti-B2GPI antibodies may improve the specificity and utility of testing.

A multicentre study of 447 patients assessed the significance of domain 1 specific IgG anti-B2GPI antibodies compared with IgG anti-B2GPI against other domains [53]. In this study, 55% of patients had IgG antibodies directed against domain 1 of B2GPI. Of these patients, 83% had a history of thrombosis. By contrast, among patients with IgG anti-B2GPI against other domains of B2GPI, a smaller proportion (58%) had a history of thrombosis. Furthermore, domain 1 specific anti-B2GPI showed a significant association with obstetric manifestations of APS in this patient group which was not observed for anti-B2GPI against other domains.

Domain specific antibodies may also be helpful in refining the diagnostic value of IgA anti-B2GPI. A study looking at a small number of SLE patients with IgA anti-B2GPI found that domain 4 and 5 specific antibodies were seen more commonly among SLE patients without thrombosis compared to those with a history of thrombosis [54]. If verified in larger cohorts of patients, domain specific anti-B2GPI antibodies could potentially help stratify patients with a higher risk of APS in association with systemic autoimmune disease which might inform clinical decisions such as when to offer anticoagulant prophylaxis. However, domain specific anti-B2GPI testing is not yet widely available in most diagnostic laboratories.

7.2. Improved Standardisation of Anti-B2GPI Testing

There are numerous challenges relating to assays used for assessment of anti-B2GPI and these are discussed extensively elsewhere [55]. These include variation in isotypes tested by different centres, variability in the sensitivity and specificity of assays, significant inter-laboratory and intra-assay variability, the lack of universally agreed units of measurement and lack of reference material for traceability of measurements.

These issues are the focus for improving the diagnostic utility of anti-B2GPI testing in APS. One area in progress is the development of IgG and IgM reference material derived from pooled serum from well characterised APS patients with very high anti-B2GPI levels. The Antiphospholipid Antibodies Task Force recently reported that further validation studies were being performed and it is hoped that this will help to address issues regarding standardisation of assays and reduce inter-laboratory and inter-assay variability [33]. There are no reference materials for IgA anti-B2GPI, and this is an area for development if IgA anti-B2GPI is to be more widely adopted in anti-B2GPI testing algorithms.

8. Conclusions

B2GPI is a multifunctional protein with key roles in the clotting pathway. Antibodies against B2GPI contribute to APS and are particularly important in the subset of patients who test negative for other aPL. Current data support the inclusion of anti-B2GPI in the laboratory diagnostic criteria of APS and in particular, favour IgG anti-B2GPI over IgM in terms of specificity for APS. There is accumulating data to support a role for IgA anti-B2GPI in APS diagnosis, but its exact significance in relation to particular APS clinical phenotypes remains unresolved. The laboratory diagnosis of APS continues to be a challenging area of research as there is considerable variability in patient populations studied, assays used and the majority of studies have been retrospective. Coordination of prospective, longitudinal studies is vital to improve our understanding of this fascinating group of autoantibodies.

Acknowledgments: The author would like to thank Jack Bourke for review of the manuscript.

Conflicts of Interest: The author declares no conflict of interest.

Abbreviations

The following abbreviations are used in this manuscript:

APS	Antiphospholipid syndrome
B2GPI	Beta-2 glycoprotein I
Anti-B2GPI	Anti-beta-2 glycoprotein I antibodies
SLE	Systemic lupus erythematosus
aPL	Antiphospholipid antibody
LA	Lupus anticoagulant
aCL	Anticardiolipin
IgG	Immunoglobulin G
IgM	Immunoglobulin M
IgA	Immunoglobulin A
ELISA	Enzyme-linked immunosorbent assay

References

1. Levine, J.S.; Branch, D.W.; Rauch, J. The antiphospholipid syndrome. *N. Eng. J. Med.* **2002**, *346*, 752–763. [CrossRef] [PubMed]
2. Cohen, D.; Berger, S.P.; Steup-Beekman, G.M.; Bloemenkamp, K.W.; Bajema, I.M. Diagnosis and management of the antiphospholipid syndrome. *BMJ* **2010**, *340*. [CrossRef] [PubMed]

3. Miyakis, S.; Lockshin, M.D.; Atsumi, T.; Branch, D.W.; Brey, R.L.; Cervera, R.; Derksen, R.H.; DE Groot, P.G.; Koike, T.; Meroni, P.L.; *et al.* International consensus statement on an update of the classification criteria for definite antiphospholipid syndrome (APS). *J. Thromb. Haemost.* **2006**, *4*, 295–306. [CrossRef] [PubMed]

4. Fischetti, F.; Durigutto, P.; Pellis, V.; Debeus, A.; Macor, P.; Bulla, R.; Bossi, F.; Ziller, F.; Sblattero, D.; Meroni, P.; *et al.* Thrombus formation induced by antibodies to beta2-glycoprotein I is complement dependent and requires a priming factor. *Blood* **2005**, *106*, 2340–2346. [CrossRef] [PubMed]

5. Meroni, P.L.; Borghi, M.O.; Raschi, E.; Tedesco, F. Pathogenesis of antiphospholipid syndrome: Understanding the antibodies. *Nat. Rev. Rheumatol.* **2011**, *7*, 330–339. [CrossRef] [PubMed]

6. De Groot, P.G.; Meijers, J.C. Beta(2)-glycoprotein I: Evolution, structure and function. *J. Thromb. Haemost.* **2011**, *9*, 1275–1284. [CrossRef] [PubMed]

7. Hammel, M.; Kriechbaum, M.; Gries, A.; Kostner, G.M.; Laggner, P.; Prassl, R. Solution structure of human and bovine beta(2)-glycoprotein I revealed by small-angle X-ray scattering. *J. Mol. Biol.* **2002**, *321*, 85–97. [CrossRef]

8. Agar, C.; van Os, G.M.; Morgelin, M.; Sprenger, R.R.; Marquart, J.A.; Urbanus, R.T.; Derksen, R.H.; Meijers, J.C.; de Groot, P.G. Beta2-glycoprotein I can exist in 2 conformations: Implications for our understanding of the antiphospholipid syndrome. *Blood* **2010**, *116*, 1336–1343. [CrossRef] [PubMed]

9. Nilsson, M.; Wasylik, S.; Morgelin, M.; Olin, A.I.; Meijers, J.C.; Derksen, R.H.; de Groot, P.G.; Herwald, H. The antibacterial activity of peptides derived from human beta-2 glycoprotein I is inhibited by protein h and m1 protein from streptococcus pyogenes. *Mol. Microbiol.* **2008**, *67*, 482–492. [CrossRef] [PubMed]

10. Agar, C.; de Groot, P.G.; Morgelin, M.; Monk, S.D.; van Os, G.; Levels, J.H.; de Laat, B.; Urbanus, R.T.; Herwald, H.; van der Poll, T.; *et al.* Beta(2)-glycoprotein I: A novel component of innate immunity. *Blood* **2011**, *117*, 6939–6947. [CrossRef] [PubMed]

11. Gropp, K.; Weber, N.; Reuter, M.; Micklisch, S.; Kopka, I.; Hallstrom, T.; Skerka, C. Beta(2)-glycoprotein I, the major target in antiphospholipid syndrome, is a special human complement regulator. *Blood* **2011**, *118*, 2774–2783. [CrossRef] [PubMed]

12. Holers, V.M.; Girardi, G.; Mo, L.; Guthridge, J.M.; Molina, H.; Pierangeli, S.S.; Espinola, R.; Xiaowei, L.E.; Mao, D.; Vialpando, C.G.; *et al.* Complement C3 activation is required for antiphospholipid antibody-induced fetal loss. *J. Exp. Med.* **2002**, *195*, 211–220. [CrossRef] [PubMed]

13. Thurman, J.M.; Kraus, D.M.; Girardi, G.; Hourcade, D.; Kang, H.J.; Royer, P.A.; Mitchell, L.M.; Giclas, P.C.; Salmon, J.; Gilkeson, G.; *et al.* A novel inhibitor of the alternative complement pathway prevents antiphospholipid antibody-induced pregnancy loss in mice. *Mol. Immunol.* **2005**, *42*, 87–97. [CrossRef] [PubMed]

14. Lonze, B.E.; Singer, A.L.; Montgomery, R.A. Eculizumab and renal transplantation in a patient with CAPS. *N. Eng. J. Med.* **2010**, *362*, 1744–1745. [CrossRef] [PubMed]

15. Wilson, W.A.; Gharavi, A.E.; Koike, T.; Lockshin, M.D.; Branch, D.W.; Piette, J.C.; Brey, R.; Derksen, R.; Harris, E.N.; Hughes, G.R.; *et al.* International consensus statement on preliminary classification criteria for definite antiphospholipid syndrome: Report of an international workshop. *Arthritis Rheum.* **1999**, *42*, 1309–1311. [CrossRef]

16. Danowski, A.; Kickler, T.S.; Petri, M. Anti-beta2-glycoprotein I: Prevalence, clinical correlations, and importance of persistent positivity in patients with antiphospholipid syndrome and systemic lupus erythematosus. *J. Rheumatol.* **2006**, *33*, 1775–1779. [PubMed]

17. Tincani, A.; Andreoli, L.; Casu, C.; Cattaneo, R.; Meroni, P. Antiphospholipid antibody profile: Implications for the evaluation and management of patients. *Lupus* **2010**, *19*, 432–435. [CrossRef] [PubMed]

18. Faden, D.; Tincani, A.; Tanzi, P.; Spatola, L.; Lojacono, A.; Tarantini, M.; Balestrieri, G. Anti-beta 2 glycoprotein I antibodies in a general obstetric population: Preliminary results on the prevalence and correlation with pregnancy outcome. Anti-beta2 glycoprotein I antibodies are associated with some obstetrical complications, mainly preeclampsia-eclampsia. *Eur. J. Obst. Gynecol. Reprpd. Biol.* **1997**, *73*, 37–42.

19. Reynaud, Q.; Lega, J.C.; Mismetti, P.; Chapelle, C.; Wahl, D.; Cathebras, P.; Laporte, S. Risk of venous and arterial thrombosis according to type of antiphospholipid antibodies in adults without systemic lupus erythematosus: A systematic review and meta-analysis. *Autoimmun. Rev.* **2014**, *13*, 595–608. [CrossRef] [PubMed]

20. Lakos, G.; Favaloro, E.J.; Harris, E.N.; Meroni, P.L.; Tincani, A.; Wong, R.C.; Pierangeli, S.S. International consensus guidelines on anticardiolipin and anti-beta2-glycoprotein I testing: Report from the 13th international congress on antiphospholipid antibodies. *Arthritis Rheum.* **2012**, *64*, 1–10. [CrossRef] [PubMed]

21. Lakos, G.; Kiss, E.; Regeczy, N.; Tarjan, P.; Soltesz, P.; Zeher, M.; Bodolay, E.; Szakony, S.; Sipka, S.; Szegedi, G. Isotype distribution and clinical relevance of anti-beta2-glycoprotein I (beta2-GPI) antibodies: Importance of IgA isotype. *Clin. Exp. Immunol.* **1999**, *117*, 574–579. [CrossRef] [PubMed]

22. Zoghlami-Rintelen, C.; Vormittag, R.; Sailer, T.; Lehr, S.; Quehenberger, P.; Rumpold, H.; Male, C.; Pabinger, I. The presence of IgG antibodies against beta2-glycoprotein I predicts the risk of thrombosis in patients with the lupus anticoagulant. *J. Thromb. Haemost.* **2005**, *3*, 1160–1165. [CrossRef] [PubMed]

23. Brusch, A.; Bundell, C.; Hollingsworth, P. Immunoglobulin G is the only anti-beta-2-glycoprotein I isotype that associates with unprovoked thrombotic events among hospital patients. *Pathology* **2014**, *46*, 234–239. [CrossRef] [PubMed]

24. Mehrani, T.; Petri, M. IgM anti-beta2 glycoprotein I is protective against lupus nephritis and renal damage in systemic lupus erythematosus. *J. Rheumatol.* **2011**, *38*, 450–453. [CrossRef] [PubMed]

25. Abinader, A.; Hanly, A.J.; Lozada, C.J. Catastrophic antiphospholipid syndrome associated with anti-beta-2-glycoprotein I IgA. *Rheumatology* **1999**, *38*, 84–85. [CrossRef]

26. Fanopoulos, D.; Teodorescu, M.R.; Varga, J.; Teodorescu, M. High frequency of abnormal levels of IgA anti-beta2-glycoprotein I antibodies in patients with systemic lupus erythematosus: Relationship with antiphospholipid syndrome. *J. Rheumatol.* **1998**, *25*, 675–680. [PubMed]

27. Yamada, H.; Tsutsumi, A.; Ichikawa, K.; Kato, E.H.; Koike, T.; Fujimoto, S. IgA-class anti-beta2-glycoprotein I in women with unexplained recurrent spontaneous abortion. *Arthritis Rheum.* **1999**, *42*, 2727–2728. [CrossRef]

28. Meijide, H.; Sciascia, S.; Sanna, G.; Khamashta, M.A.; Bertolaccini, M.L. The clinical relevance of IgA anticardiolipin and IgA anti-beta2 glycoprotein I antiphospholipid antibodies: A systematic review. *Autoimmun. Rev.* **2013**, *12*, 421–425. [CrossRef] [PubMed]

29. Samarkos, M.; Davies, K.A.; Gordon, C.; Loizou, S. Clinical significance of IgA anticardiolipin and anti-beta2-gp1 antibodies in patients with systemic lupus erythematosus and primary antiphospholipid syndrome. *Clin. Rheumatol.* **2006**, *25*, 199–204. [CrossRef] [PubMed]

30. Murthy, V.; Willis, R.; Romay-Penabad, Z.; Ruiz-Limon, P.; Martinez-Martinez, L.A.; Jatwani, S.; Jajoria, P.; Seif, A.; Alarcon, G.S.; Papalardo, E.; *et al.* Value of isolated IgA anti-beta2 -glycoprotein I positivity in the diagnosis of the antiphospholipid syndrome. *Arthritis Rheum.* **2013**, *65*, 3186–3193. [CrossRef] [PubMed]

31. Ruiz-Garcia, R.; Serrano, M.; Martinez-Flores, J.A.; Mora, S.; Morillas, L.; Martin-Mola, M.A.; Morales, J.M.; Paz-Artal, E.; Serrano, A. Isolated IgA anti-beta2 glycoprotein I antibodies in patients with clinical criteria for antiphospholipid syndrome. *J. Immunol. Res.* **2014**, *2014*. [CrossRef]

32. Martinez-Flores, J.A.; Serrano, M.; Alfaro, J.; Mora, S.; Paz-Artal, E.; Morales, J.M.; Serrano, A. Heterogeneity between diagnostic tests for IgA anti-beta2 glycoprotein I: Explaining the controversy in studies of association with vascular pathology. *Anal. Chem.* **2013**, *85*, 12093–12098. [CrossRef] [PubMed]

33. Bertolaccini, M.L.; Amengual, O.; Andreoli, L.; Atsumi, T.; Chighizola, C.B.; Forastiero, R.; de Groot, P.; Lakos, G.; Lambert, M.; Meroni, P.; *et al.* 14th international congress on antiphospholipid antibodies task force. Report on antiphospholipid syndrome laboratory diagnostics and trends. *Autoimmun. Rev.* **2014**, *13*, 917–930. [CrossRef] [PubMed]

34. Asherson, R.A.; Cervera, R. Antiphospholipid antibodies and infections. *Ann. Rheum. Dis.* **2003**, *62*, 388–393. [CrossRef] [PubMed]

35. Loizou, S.; Singh, S.; Wypkema, E.; Asherson, R.A. Anticardiolipin, anti-beta(2)-glycoprotein I and antiprothrombin antibodies in black south African patients with infectious disease. *Ann. Rheum Dis.* **2003**, *62*, 1106–1111. [CrossRef] [PubMed]

36. Gabeta, S.; Norman, G.L.; Gatselis, N.; Liaskos, C.; Papamichalis, P.A.; Garagounis, A.; Zachou, K.; Rigopoulou, E.I.; Dalekos, G.N. IgA anti-B2GPI antibodies in patients with autoimmune liver diseases. *J. Clin. Immunol.* **2008**, *28*, 501–511. [CrossRef] [PubMed]

37. Mankai, A.; Achour, A.; Thabet, Y.; Manoubia, W.; Sakly, W.; Ghedira, I. Anti-cardiolipin and anti-beta 2-glycoprotein I antibodies in celiac disease. *Pathol. Biol.* **2012**, *60*, 291–295. [CrossRef] [PubMed]

38. Borges, R.B.; Bodanese, L.C.; Muhlen, C.A.; Repetto, G.; Viehe, M.; Norman, G.L.; Staub, H.L. Anti-beta2-glycoprotein I autoantibodies and metabolic syndrome. *Arq. Bras. Cardiol.* **2011**, *96*, 272–276. [CrossRef] [PubMed]

39. Serrano, A.; Garcia, F.; Serrano, M.; Ramirez, E.; Alfaro, F.J.; Lora, D.; de la Camara, A.G.; Paz-Artal, E.; Praga, M.; Morales, J.M. IgA antibodies against beta2 glycoprotein I in hemodialysis patients are an independent risk factor for mortality. *Kidney Int.* **2012**, *81*, 1239–1244. [CrossRef] [PubMed]

40. Serrano, M.; Martinez-Flores, J.A.; Castro, M.J.; Garcia, F.; Lora, D.; Perez, D.; Gonzalez, E.; Paz-Artal, E.; Morales, J.M.; Serrano, A. Renal transplantation dramatically reduces IgA anti-beta-2-glycoprotein I antibodies in patients with endstage renal disease. *J. Immunol. Res.* **2014**, *2014*. [CrossRef] [PubMed]

41. De Jesus, G.R.; Agmon-Levin, N.; Andrade, C.A.; Andreoli, L.; Chighizola, C.B.; Porter, T.F.; Salmon, J.; Silver, R.M.; Tincani, A.; Branch, D.W. 14th international congress on antiphospholipid antibodies task force report on obstetric antiphospholipid syndrome. *Autoimmun. Rev.* **2014**, *13*, 795–813. [CrossRef] [PubMed]

42. Abou-Nassar, K.; Carrier, M.; Ramsay, T.; Rodger, M.A. The association between antiphospholipid antibodies and placenta mediated complications: A systematic review and meta-analysis. *Thromb. Res.* **2011**, *128*, 77–85. [CrossRef] [PubMed]

43. Katano, K.; Aoki, A.; Sasa, H.; Ogasawara, M.; Matsuura, E.; Yagami, Y. Beta 2-glycoprotein I-dependent anticardiolipin antibodies as a predictor of adverse pregnancy outcomes in healthy pregnant women. *Hum. Reprod.* **1996**, *11*, 509–512. [CrossRef] [PubMed]

44. Lee, R.M.; Brown, M.A.; Branch, D.W.; Ward, K.; Silver, R.M. Anticardiolipin and anti-beta2-glycoprotein-i antibodies in preeclampsia. *Obstet. Gynecol.* **2003**, *102*, 294–300. [PubMed]

45. Lee, R.M.; Emlen, W.; Scott, J.R.; Branch, D.W.; Silver, R.M. Anti-beta-2-glycoprotein I antibodies in women with recurrent spontaneous abortion, unexplained fetal death, and antiphospholipid syndrome. *Am. J. Obstet. Gynecol.* **1999**, *181*, 642–648. [CrossRef]

46. Valdes-Macho, E.; Cabiedes, J.; Villa, A.R.; Cabral, A.R.; Alarcon-Segovia, D. Anticardiolipin and anti-beta2-glycoprotein-i antibodies in hypertensive disorders of pregnancy. *Arch. Med. Res.* **2002**, *33*, 460–465. [CrossRef]

47. Vora, S.; Shetty, S.; Salvi, V.; Satoskar, P.; Ghosh, K. A comprehensive screening analysis of antiphospholipid antibodies in Indian women with fetal loss. *Eur. J. Obst. Gynecol. Rep. Biol.* **2008**, *137*, 136–140. [CrossRef] [PubMed]

48. Silver, R.M.; Parker, C.B.; Reddy, U.M.; Goldenberg, R.; Coustan, D.; Dudley, D.J.; Saade, G.R.; Stoll, B.; Koch, M.A.; Conway, D.; *et al.* Antiphospholipid antibodies in stillbirth. *Obstet. Gynecol.* **2013**, *122*, 641–657. [CrossRef] [PubMed]

49. Ruffatti, A.; Tonello, M.; Visentin, M.S.; Bontadi, A.; Hoxha, A.; De Carolis, S.; Botta, A.; Salvi, S.; Nuzzo, M.; Rovere-Querini, P.; *et al.* Risk factors for pregnancy failure in patients with anti-phospholipid syndrome treated with conventional therapies: A multicentre, case-control study. *Rheumatology* **2011**, *50*, 1684–1689. [CrossRef] [PubMed]

50. Ulcova-Gallova, Z.; Bouse, V.; Krizanovska, K.; Balvin, M.; Rokyta, Z.; Netrvalova, L. Beta 2-glycoprotein I is a good indicator of certain adverse pregnancy conditions. *Int. J. Fertil. Womens Med.* **2001**, *46*, 304–308. [PubMed]

51. Lee, R.M.; Branch, D.W.; Silver, R.M. Immunoglobulin A anti-beta2-glycoprotein antibodies in women who experience unexplained recurrent spontaneous abortion and unexplained fetal death. *Am. J. Obstet. Gynecol.* **2001**, *185*, 748–753. [CrossRef] [PubMed]

52. Carmo-Pereira, S.; Bertolaccini, M.L.; Escudero-Contreras, A.; Khamashta, M.A.; Hughes, G.R. Value of IgA anticardiolipin and anti-beta2-glycoprotein I antibody testing in patients with pregnancy morbidity. *Ann. Rheum. Dis.* **2003**, *62*, 540–543. [CrossRef] [PubMed]

53. De Laat, B.; Pengo, V.; Pabinger, I.; Musial, J.; Voskuyl, A.E.; Bultink, I.E.; Ruffatti, A.; Rozman, B.; Kveder, T.; de Moerloose, P.; *et al.* The association between circulating antibodies against domain 1 of beta2-glycoprotein I and thrombosis: An international multicenter study. *J. Thromb. Haemost.* **2009**, *7*, 1767–1773. [CrossRef] [PubMed]

54. Despierres, L.; Beziane, A.; Kaplanski, G.; Granel, B.; Serratrice, J.; Cohen, W.; Bretelle, F.; Rossi, P.; Morange, P.E.; Weiller, P.J.; *et al.* Contribution of anti-beta2glycoprotein I IgA antibodies to the diagnosis of anti-phospholipid syndrome: Potential interest of target domains to discriminate thrombotic and non-thrombotic patients. *Rheumatology* **2014**, *53*, 1215–1218. [CrossRef] [PubMed]

55. Favaloro, E.J.; Wong, R.C. Antiphospholipid antibody testing for the antiphospholipid syndrome: A comprehensive practical review including a synopsis of challenges and recent guidelines. *Pathology* **2014**, *46*, 481–495. [CrossRef] [PubMed]

Review

Antiphospholipid Syndrome and Kidney Involvement: New Insights

José A. Martínez-Flores [1], Manuel Serrano [1], Jose M. Morales [1] and Antonio Serrano [1,2,*]

[1]   Department of Immunology, Instituto de Investigación Hospital Universitario 12 de Octubre, 28041 Madrid, Spain; martinezfloresja@gmail.com (J.A.M.-F.); mserranobl@gmail.com (M.S.); jmorales@h12o.es (J.M.M.)
[2]   Immunology Section, Universidad San Pablo-CEU, 28660 Madrid, Spain
*   Correspondence: aserrano@h12o.es; Tel.: +34-91-779-2602

Academic Editor: Ricard Cervera
Received: 18 May 2016; Accepted: 1 July 2016; Published: 11 July 2016

Abstract: Antiphospholipid syndrome is an autoimmune disorder characterized by vascular thromboses and pregnancy morbidity associated with antiphospholipid antibodies: lupus anticoagulant, IgG or IgM anticardiolipin or anti-beta 2-glycoprotein I. The kidney is one of the major target organs in antiphospholipid syndrome (APS). However, beyond the known involvement of the kidney in primary and associated APS, we may be observing a new form of APS within the context of renal failure. This review describes the classical kidney manifestations of APS and provides new considerations to be taken into account.

Keywords: autoimmunity; antiphospholipid antibodies; end stage kidney disease; early graft loss; thrombosis antiphospholipid syndrome; aPL; B2GPI; IgA

1. Introduction

Hughes syndrome, most commonly called antiphospholipid syndrome (APS), is an autoimmune multisystemic disorder characterized by thrombosis and pregnancy-related complications in patients with antiphospholipid antibodies (aPL) [1]. aPL comprise a heterogeneous group of autoantibodies directed against phospholipids or protein–phospholipids complexes [2,3].

The diagnosis of APS is made when one clinical and at least one laboratory criteria are fulfilled [4]. The international consensus statement updated in 2006 includes as clinical criteria one or more episodes of thrombosis, arterial or venous, in any tissue or organ. Clinical criteria must be confirmed by unequivocal techniques such as imaging studies or histopathology. Consensus APS pregnancy morbidity includes: (1) unexplained fetus death at or beyond the 10th week of gestation; (2) premature birth before the 34th week of gestation due to eclampsia, preeclampsia or placental insufficiency; and (3) three or more unexplained abortions before the 10th week of gestation.

Lupus anticoagulant (LA) in plasma, IgG and/or IgM anticardiolipin (aCL) in medium or higher titers (>40 GPL or MPL, or > the 99th percentile) and anti-beta-2-glycoprotein I (aB2GPI) antibodies (in titer > the 99th percentile) in serum or plasma determined in at least two occasions separated by at least 12 weeks comprises the laboratory criteria [4].

APS can be classified into four groups in accordance with the type of aPL presented by the patient [4]:

I—More than one laboratory criterion in any combination.

II a—LA present isolated.

II b—aCL IgG or IgM isolated.

II c—aB2PGI isolated IgG or IgM.

Although anti-B2GPI antibodies of IgA isotype were not included in the 2004-defined laboratory criteria for APS due to controversial results [5], researchers were encouraged to clarify their role in the APS in the same meeting [4]. The clinical importance of IgA aB2GPI has increased in the last few years [6,7] due to the utilization of kits useful to detect IgA aB2GPI [8] and the task force in the 13th International Congress on Antiphospholipid Antibodies (2010, Galveston, TX, USA) recommended testing for the IgA aB2GPI in cases negative for IgG and IgM and when APS is still suspected [9,10].

APS was first described by Hughes in the mid-1980s as a disorder of hypercoagulability in association with aPL [11]. This syndrome was first reported in a patient with systemic lupus erythematosus (SLE) [12]. However, shortly after APS was defined in this way, several authors suggested separate categories in order to group patients with APS. At present, there are three different APS disease entities: (1) Patients with APS criteria and without associated systemic autoimmune disorders are defined as having "primary antiphospholipid syndrome" (PAPS) [13,14]. This is currently the most common form of the disease [15]; (2) Patients with associated systemic autoimmune disorders such as SLE or rheumatoid arthritis (RA) are classified as having "associated antiphospholipid syndrome" (SAPS) [13,14]; (3) Patients with catastrophic antiphospholipid syndrome (CAPS), consisting in multiple organ thrombosis with simultaneous multiorgan failure and mortality rate near to 50% [14].

Although prevalence of aPL in healthy populations is between 1.0% [16] and 5.6% [17], the real prevalence of APS among the general population is unknown. The prevalence published in different works show significant variability [18] and APS is probably underdiagnosed. aPL are not exclusive to autoimmune diseases as they have also been detected in infections [19,20], neoplasias [21] or are chlorpromazine [22] or quinidine induced [23]. The estimated incidence of APS is approximately 5 cases per 100,000 persons per year [17].

It is unclear how thrombosis is developed in aPL. Some studies have associated aPL with different thrombotic pathways, including persistent activation of coagulation [24,25], platelet activation [26], vessel inflammation [27] or accelerated atherosclerosis [28].

Nevertheless, the presence of aPL is not the only condition to induce thrombosis formation. Patients with aPL for long periods of time need a "second hit" involving activation of innate immunity and a pro-inflammatory microenvironment to trigger thrombotic episodes. Infections or surgery have been suggested as a possible means of a second hit [14,29].

Establishment of consensus criteria for APS has made it possible to standardize patient groups but it has also generated controversy because these very restrictive criteria have limited use for clinical purposes [30]. Some authors have proposed a redefinition of APS [31]. In fact, manifestations associated with aPL antibodies such as heart valve disease, livedo reticularis (LR), some neurological manifestations such as migraines or epilepsy, stroke or thrombocytopenia were not included in the updated criteria despite some of them being frequent in APS patients [18,32].

Renal manifestations are one of the most prevalent events in APS patients and may co-exist with other conditions, especially with lupus nephritis. Renal involvement in APS is caused by thrombosis in any location within the kidney vasculature. Renal manifestations may include renal infarction, renovascular hypertension, APS nephropathy, thrombotic microangiopathy (TMA), renal artery stenosis and increased allograft thrombosis [33].

The true incidence of renal involvement in APS has not been well determined and may be underdiagnosed. This involvement may fluctuate between 9% and 10% of patients having APS [34]. The purpose of this review is to summarize the information about APS and renal involvement.

2. APS Nephropathy

APS nephropathy is clinically characterized by a syndrome of vascular nephropathy associated with proteinuria, hypertension and renal insufficiency [35]. Acute presentation of APS nephropathy includes TMA. TMA has distinctive patterns and is usually associated with nephortic range proteinuria [36,37], hypertension and renal failure [38]. aPL test is always a useful tool to distinguish whether TMA is caused by APS [38] instead of other entities such as hemolitic uremic syndrome or thtombotic thrombocitopenic purpura [39].

The most striking features of APS nephropathy are vasocclusive lesions (total or partial), including TMA [37]. Focal or diffuse microangiopathic areas that might affect any of the vessels within the renal vasculature, recanalizing thrombi in arteries and arterioles are other well-defined characteristics [36]. Nevertheless, other changes in biopsies have been described unrelated to thrombosis such as glomerular basement membrane reduplication, focal segmental glomerulosclerosis, intimal hyperplasia of arteries or acute tubular necrosis [36,37,40,41].

3. Systemic Hypertension

Hypertension associated with livedo reticularis was described by Hughes in the definition of APS. Systemic hypertension is considered a sensitive marker of nephropathy. Nochy et al. reported a prevalence of 93%, in some cases this being the only clinical sign suggesting nephropathy [37]. Cacoub et al. reported malignant hypertension in APS patients without SLE in five patients [42].

4. Renal Artery Lesions

Renal artery involvement can be uni- or bilateral and usually consists of occlusive lesions resulting from thrombosis. The most common clinical manifestation is renal thrombosis in the onset of severe hypertension [43].

Renal artery stenosis (RAS) is another characteristic condition in APS. A total of 26% of aPL positive patients with non-controlled hypertension were found to have renal artery stenosis, this being significantly higher than in hypertensive controls (8%) and healthy potential kidney donors [43]. Although the source of RAS is not clear, the response to anticoagulation suggests a thrombotic process [43,44].

5. Renal Vein Thrombosis

Thrombosis may occur in the main and/or minor renal veins of patients with PAPS and patients with associated SLE [45,46]. This manifestation has been particularly associated with LA positivity and is usually characterized by the presence of nephrotic range proteinuria [35,46].

6. Lupus Nephritis

APS nephropathy in SLE patients may vary from 11.4% to 39.5% according to several studies [47]. Silvariño et al. found that patients with LA plus IgG aCL had an increased prevalence of APS nephropathy, suggesting that these aPL may have a direct effect on the development of these lesions [48]. Furthermore, Zheng et al. found that LA accompanied by antibodies against B2GPI is related to glomerular microthrombosis, with a prevalence ranging from 20% to 30% [49].

7. Catastrophic Antiphospholipid Syndrome (CAPS)

CAPS is an accelerated variant of APS with multiorgan failure being described for the first time by Asherson [50,51]. Although many patients with APS have venous or arterial thrombosis within an isolated location in the body, a small proportion have undergone rapid onset of multiorgan thrombosis associated with high mortality [52].

Due to the low prevalence of this form of APS, estimated at 0.8%, there is little clinical experience regarding this form [15]. At least three organs must be affected in order to diagnose the CAPS form, and symptoms may develop within a few days or weeks. Although large vessels may be affected, more commonly there is thrombotic microangiopathy affecting small vessels in multiple organs. In 50% of cases, the heart, central nervous system, liver, skin and kidney are affected [53]. According to the CAPS registry data, at least 70% of the patients had renal involvement [54]. Death is usually the result of multiple organ failure in more than 50% in patients with APS [52,55].

It is noteworthy that approximately 60% of the CAPS are preceded by a precipitating event, mainly infections [35].

8. aPL in Hemodialysis Patients

End stage renal disease is a rare complication of APS and only a few studies have investigated this relationship. High prevalence of aPL in hemodialysis (Table 1) has been reported. These antibodies seem to be independent of age, sex, length of time on dialysis, drugs and chronic B and C hepatitis [56]. The mechanism by which these autoantibodies appear is not clear. However, dialysis membranes, trauma to blood in dialysis or infections have been postulated as possible causes [56]. Cardiovascular complications rank first among the causes of death in patients on dialysis with chronic kidney disease and end-stage renal disease and include thrombotic episodes in cerebral and myocardial vessels [57]. The presence of these autoantibodies is unclear because some studies have associated aPL with thrombotic events [58,59] but other studies have not [60,61]. Overall, the evidence with classic aPL is not clear regarding the pathogenic role in end stage renal disease.

Table 1. Prevalence of consensus aPL in hemodialysis patients.

Reference	Patients	LA (%)	aCL (%)	aB2GPI (%)
Quereda (Nephron, 1988) [62]	56	22		
Grönhagen (BMJ, 1990) [63]	146		26	
García- Martín (NDT, 1991) [33]	51	22	31	
Matsuda (Thr Res, 1993) [64]	39	33	31	
Brunet (Kidney Int, 1995) [56]	97	19	15.5	
Valeri (Clin nephrol, 1996) [65]	230		29	
Ducloux (Kidney Int, 2003) [66]	324		26.8	
García-Martín (doctoral tesis, 1999)	192	26	36	18
TOTAL	939	18.7	26	18

9. aPL in Renal Transplantation

aPL in renal transplantation varies from 2% [34,67] up to 15% [68] of the patients. Sufficient evidence exists to show that aPL positive patients have a higher risk of developing thrombosis with consequent graft loss. Vaidya et al. reported six patients positive for aPL with renal graft loss by thrombosis in the first week of the transplantation [69]. Wagenknecht et al. found that 57% of patients with primary nonfunctioning renal allografts were aPL positive, the latter posing an important risk factor for early renal allograft failure [70]. However, some studies have not reported any association between aPL and risk of graft loss despite a high prevalence of pretransplant aPL [68]. The small number of patients included in the study may be the key to the non-association of aPL with graft loss.

It has also been reported that a minority of patients (15.7% of patients with aPL) can develop aPL post-transplantation. It is still unknown how patients can acquire aPL but it has been speculated that it may be caused by a post-transplantation infection [71]. Furthermore, the development of aPL after transplantation was associated with acute rejection [72].

10. The Role of IgA aB2GPI Antibodies in the Context of End Stage Kidney Disease

10.1. Dialysis

Patients with chronic renal failure have a high prevalence of cardiovascular problems, which are especially severe in patients on dialysis [73,74]. Cardiovascular complications are ranked first in cause of death in dialysis patients including thrombotic events in the brain and myocardium [57]. Endothelial dysfunctions are common in these patients and this may contribute to hypertension and arteriosclerosis [75].

As described previously, the use of IgA aB2GPI is not an accepted laboratory criterion for the diagnosis of APS. However, since Galveston 2010 IgA, aB2GPI testing has been recommended in patients with persistent clinical of APS and negative for consensus aPL [76].

In a prospective study, 124 patients on dialysis were followed-up for two years. Of these, 41 (33.1%) were positive for IgA aB2GPI and 83 negative. Of the 41 positive, 46.3% died during the follow-up while only 16.9% of the 83 negative patients died ($p < 0.001$). The main cause of death for IgA aB2GPI positive patients was cardiovascular. Survival in the group of IgA aB2GPI positive patients was significantly lower than in the negatives. The risk of death (Hazard Ratio) for IgA aB2GPI positive patients was 3.274 (for a confidence interval of 95%, $p < 0.001$) [77].

The prevalence of consensus aPL was aCL IgG (9.7%), aCL IgM (6.4%) and aB2GPI IgG (10.9%), however, consensus aPL were not related to mortality or vascular events. It has been speculated that antigens released from the membranes of endothelial cells interact with dialysis membranes complexed hapten, exposing new epitopes in hemodialysis patients [77]. These complexes could stimulate the formation of aB2GPI antibodies that can promote hypercoagulability and arteriosclerosis progression [78].

Another study with patients undergoing hemodialysis recently confirmed the high prevalence of IgA aB2GPI antibodies and their relationship to vascular complications [79].

Serrano et al. described a similar prevalence of IgA aB2GPI in patients undergoing peritoneal and hemodialysis, indicating that the proposed models of hapten-carriers complexes generated by B2GPI interaction with dialysis membranes and endothelial injury by dialysis system access to body could be discarded [80].

10.2. Renal Transplantation

Classic aPL in transplantation are scarce, however they seem to be a clear association with graft loss. Two studies have investigated the implications of IgA aB2GPI in renal transplantation. A 10-year follow-up prospective study was performed including all patients transplanted during the years 2000–2002 in the 12 Octubre Hospital. Pretransplant IgA aB2GPI was examined retrospectively in 269 patients. Eighty-nine patients (33%, group 1) were positive for IgA aB2GPI and 180 patients (67%, group 2) were negative. Group 1 had a higher incidence of early graft loss at six months (11.2% in Group 1 vs. 1.6% in Group 2; $p = 0.002$) the thrombosis of the vessels being the most prevalent cause of graft loss and was only observed in Group 1. In the multivariate analysis, IgA aB2GPI was an independent risk factor for the development of early graft loss and delayed graft function but was unable to prove IgA aB2GPI as an independent risk factor for thrombosis graft loss. Graft survival was similar in both groups after the first six months. Therefore, patients positive for IgA aB2GPI pretransplant have a high risk of early graft loss and a high risk of delayed graft function [41]. An extension of this study including 1375 patients made possible to relate IgA aB2GPI as an independent predictor of early graft loss due to thrombosis [81].

IgA aB2GPI antibodies drop immediately after renal transplantation and remain significantly lower than the levels observed in the pre-transplant status, even in patients who have lost their graft and have returned to dialysis [80]. This fact leads to a therapeutic doubt on whether immunosuppression can be effective in patients with APS, and suggests that further study is necessary.

How IgA aB2GPI induce early graft loss due to thrombosis is not clear but it has been speculated with surgery as a possible "second hit" triggering thrombosis [81]. This should be investigated by other groups because only one group has been described this relationship.

11. Treatment

There is no specific treatment for APS. Treatments administered to APS patients are primarily preventive. They consist in interfering with clotting mechanisms to prevent further thrombotic events. The drugs most used are unfractionated heparin (UFH), low molecular weight heparin (LMWH) and vitamin K antagonists (VKA).

UFH or LMWH are used in acute thrombotic events [82]. VKA anticoagulation is the treatment of choice in the long term, especially in patients with venous thromboembolism and embolic cerebral ischemia [83]. Treatment intensity is measured by the International Normalized Ratio (INR), which should range from 2.0 to 3.0 in these cases. The same treatment with or without low dose aspirin is indicated for arterial thrombosis [83]. Aspirin is indicated for patients with previous pregnancy morbidity with no thrombotic event. Results from observational studies have indicated a protective effect of aspirin in patients with asymptomatic SLE [84–87] and 197 women with APS [88]. Catastrophic APS should be treated with a combination of VKA, low dose aspirin and/or intravenous and/or plasma immunoglobulins [89]. Rituximab, hydroxycloroquine, plasma exchange, anti-C5 antibodies or statins can also be an alternative treatment for patients persistently positive for aPL [83,90]. Eculizumab is a complement inhibitor that has shown efficacy in preventing the recurrence of APS events after renal transplantation [91].

The development of new direct oral anticoagulant (DOAC) drugs has enabled new treatments. Dabigatran is a DOAC that has been shown to be effective in arterial thrombosis treatment [92]. Rivaroxaban is a selective inhibitor of factor Xa and has no effect on platelet activation [93]. Fondaparinux is a selective inhibitor of factor Xa. It has been used as a treatment for heparin-induced thrombocytopenia, but recently it has been found that Fondaparinux prevents binding of the B2GPI antibody to the B2GPI [94]. These anticoagulants do not require monitoring, which is a greater advance on anticoagulation. Nevertheless, because of its high price and its similar efficacy [93,95] with the current anticoagulants, it plays a secondary role.

12. New Perspectives

Despite treatments for APS, a multicenter prospective study of 1000 APS patients followed-up for 10 years concluded that it was necessary to search for new markers to prevent the complications of APS. Even though patients were under treatment, some of them continued to develop thrombosis [96]. Recently, a new method to detect specific circulating immune complexes (CIC) of IgA bounded to B2GPI (B2A-CIC) has been described [97]. The B2A-CIC identifies a subgroup of patients prone to develop thrombosis [98]. Testing of this B2A-CIC in the pretransplant population could provide a way to identify patients prone to graft loss. Other aPL such as anti-phosphatidilserine/prothrombin [99] or anti-annexin [100] may be helpful to identify new patients to be treated. It has been recently described that the mTORC pathway is involved in vascular lesions associated with the APS [101]. In this way, initial immunosuppression based on mTOR-inhibitors in renal transplant patients positive for IgA-aB2GPI-ab could be an alternative to the use of VKA to prevent thrombotic events.

13. Conclusions

APS is well recognized as an important cause of kidney injury, with specific clinical and histological features that may lead to renal injury caused by thrombosis at any location within the renal vasculature. Prompt evaluation should be performed in APS patients with manifestations such as systemic hypertension, livedo reticularis, hematuria, proteinuria or renal insufficiency without any other justifying etiology. Testing for aPL must also be considered for patients with any of these manifestations, especially IgA aB2GPI antibodies, the latter representing about 30% of patients in dialysis and transplantation. APS has focused mainly on patients with SLE, but patients with SAPS and renal failure only represents 2%–5% in hemodialysis or transplantation. We may be witnessing a new form of APS based on IgA aB2GPI antibodies not related with autoimmune diseases, which should be studied in the upcoming years. For patients with SLE and positive for aPL, APS nephropathy, alone or associated with SLE nephritis, should be considered in order to help guide prompt therapeutic decisions that may help to prevent the development of renal failure.

Anticoagulation remains the main treatment for patients with renal disease caused by APS. Future studies may help to identify better therapeutic targets.

Antibodies **2016**, *5*, 17

Acknowledgments: We thank Barbara Shapiro for her excellent work of translation and English revision of the paper. This work was supported by grants from Fondo de Investigaciones Sanitarias and cofinanced by European Regional Development Fund (Grants: PI12-0108, PIE13/0045 and PI14-0360).

Author Contributions: All authors designed the review and wrote the manuscript.

Conflicts of Interest: The authors declare no conflict of interest.

References

1. Willis, R.; Harris, E.N.; Pierangeli, S.S. Pathogenesis of the antiphospholipid syndrome. *Semin. Thromb. Hemost.* **2012**, *38*, 305–321. [CrossRef] [PubMed]
2. Triplett, D.A. Antiphospholipid antibodies. *Arch. Pathol. Lab. Med.* **2002**, *126*, 1424–1429. [PubMed]
3. Fischer, M.J.; Rauch, J.; Levine, J.S. The antiphospholipid syndrome. *Semin. Nephrol.* **2007**, *27*, 35–46. [CrossRef] [PubMed]
4. Miyakis, S.; Lockshin, M.D.; Atsumi, T.; Branch, D.W.; Brey, R.L.; Cervera, R.; Derksen, R.H.W.M.; de Groot, P.G.; Koike, T.; Meroni, P.L.; et al. International consensus statement on an update of the classification criteria for definite antiphospholipid syndrome (APS). *J. Thromb. Haemost.* **2006**, *4*, 295–306. [CrossRef] [PubMed]
5. Branch, D.W. Summary of the 11th international congress on antiphospholipid autoantibodies, australia, november 2004. *J. Reprod. Immunol.* **2005**, *66*, 85–90. [CrossRef] [PubMed]
6. Ruiz-Garcia, R.; Serrano, M.; Martinez-Flores, J.A.; Mora, S.; Morillas, L.; Martin-Mola, M.A.; Morales, J.M.; Paz-Artal, E.; Serrano, A. Isolated IgA anti-beta2-glycoprotein I antibodies in patients with clinical criteria for antiphospholipid syndrome. *J. Immunol. Res.* **2014**, *2014*. [CrossRef] [PubMed]
7. Murthy, V.; Willis, R.; Romay-Penabad, Z.; Ruiz-Limon, P.; Martinez-Martinez, L.A.; Jatwani, S.; Jajoria, P.; Seif, A.; Alarcon, G.S.; Papalardo, E.; et al. Value of isolated IgA anti-beta2-glycoprotein I positivity in the diagnosis of the antiphospholipid syndrome. *Arthritis Rheumatol.* **2013**, *65*, 3186–3193. [CrossRef] [PubMed]
8. Martinez-Flores, J.A.; Serrano, M.; Alfaro, J.; Mora, S.; Paz-Artal, E.; Morales, J.M.; Serrano, A. Heterogeneity between diagnostic tests for IgA anti-beta2-glycoprotein I: Explaining the controversy in studies of association with vascular pathology. *Anal. Chem.* **2013**, *85*, 12093–12098. [CrossRef] [PubMed]
9. Lakos, G.; Favaloro, E.J.; Harris, E.N.; Meroni, P.L.; Tincani, A.; Wong, R.C.; Pierangeli, S.S. International consensus guidelines on anticardiolipin and anti-beta2-glycoprotein I testing: Report from the 13th international congress on antiphospholipid antibodies. *Arthritis Rheumatol.* **2012**, *64*, 1–10. [CrossRef] [PubMed]
10. Meijide, H.; Sciascia, S.; Sanna, G.; Khamashta, M.A.; Bertolaccini, M.L. The clinical relevance of IgA anticardiolipin and IgA anti-beta2-glycoprotein I antiphospholipid antibodies: A systematic review. *Autoimmun. Rev.* **2013**, *12*, 421–425. [CrossRef] [PubMed]
11. Hughes, G.R. Thrombosis, abortion, cerebral disease, and the lupus anticoagulant. *Br. Med. J. (Clin. Res. Ed.)* **1983**, *287*, 1088–1089. [CrossRef]
12. Glueck, H.I.; Kant, K.S.; Weiss, M.A.; Pollak, V.E.; Miller, M.A.; Coots, M. Thrombosis in systemic lupus erythematosus. Relation to the presence of circulating anticoagulants. *Arch. Intern. Med.* **1985**, *145*, 1389–1395. [CrossRef] [PubMed]
13. Alarcon-Segovia, D.; Sanchez-Guerrero, J. Primary antiphospholipid syndrome. *J. Rheumatol.* **1989**, *16*, 482–488. [PubMed]
14. Harris, E.N.; Pierangeli, S.S. Primary, secondary, and catastrophic antiphospholipid syndrome: What's in a name? *Semin. Thromb. Hemost.* **2008**, *34*, 219–226. [CrossRef] [PubMed]
15. Cervera, R.; Piette, J.C.; Font, J.; Khamashta, M.A.; Shoenfeld, Y.; Camps, M.T.; Jacobsen, S.; Lakos, G.; Tincani, A.; Kontopoulou-Griva, I.; et al. Antiphospholipid syndrome: Clinical and immunologic manifestations and patterns of disease expression in a cohort of 1000 patients. *Arthritis Rheumatol.* **2002**, *46*, 1019–1027. [CrossRef] [PubMed]
16. Vaarala, O.; Palosuo, T.; Kleemola, M.; Aho, K. Anticardiolipin response in acute infections. *Clin. Immunol. Immunopathol.* **1986**, *41*, 8–15. [CrossRef]
17. Gomez-Puerta, J.A.; Cervera, R. Diagnosis and classification of the antiphospholipid syndrome. *J. Autoimmun.* **2014**, *48–49*, 20–25. [CrossRef] [PubMed]

18. Biggioggero, M.; Meroni, P.L. The geoepidemiology of the antiphospholipid antibody syndrome. *Autoimmun. Rev.* **2010**, *9*, A299–A304. [CrossRef] [PubMed]
19. Blank, M.; Krause, I.; Fridkin, M.; Keller, N.; Kopolovic, J.; Goldberg, I.; Tobar, A.; Shoenfeld, Y. Bacterial induction of autoantibodies to beta2-glycoprotein-I accounts for the infectious etiology of antiphospholipid syndrome. *J. Clin. Investig.* **2002**, *109*, 797–804. [CrossRef] [PubMed]
20. Pellicano, R.; Broutet, N.; Ponzetto, A.; Megraud, F. Helicobacter pylori: From the stomach to the heart. *Eur. J. Gastroenterol. Hepatol.* **1999**, *11*, 1335–1337. [CrossRef] [PubMed]
21. Harel, M.; Aron-Maor, A.; Sherer, Y.; Blank, M.; Shoenfeld, Y. The infectious etiology of the antiphospholipid syndrome: Links between infection and autoimmunity. *Immunobiology* **2005**, *210*, 743–747. [CrossRef] [PubMed]
22. Ducloux, D.; Florea, A.; Fournier, V.; Rebibou, J.M.; Chalopin, J.M. Inferior vena cava thrombosis in a patient with chlorpromazin-induced anticardiolipin antibodies. *Nephrol. Dial Transplant.* **1999**, *14*, 1335–1336. [CrossRef] [PubMed]
23. Clauser, S.; Fischer, A.M.; Darnige, L. Quinidine-induced lupus anticoagulant, hypoprothrombinemia, and antiprothrombin antibodies. *Am. J. Hematol.* **2007**, *82*, 330. [CrossRef] [PubMed]
24. Ieko, M.; Ichikawa, K.; Atsumi, T.; Takeuchi, R.; Sawada, K.I.; Yasukouchi, T.; Koike, T. Effects of beta2-glycoprotein I and monoclonal anticardiolipin antibodies on extrinsic fibrinolysis. *Semin. Thromb. Hemost.* **2000**, *26*, 85–90. [CrossRef] [PubMed]
25. McNeil, H.P.; Simpson, R.J.; Chesterman, C.N.; Krilis, S.A. Anti-phospholipid antibodies are directed against a complex antigen that includes a lipid-binding inhibitor of coagulation: Beta2-glycoprotein I (apolipoprotein h). *Proc. Natl. Acad. Sci. USA* **1990**, *87*, 4120–4124. [CrossRef] [PubMed]
26. Raschi, E.; Testoni, C.; Bosisio, D.; Borghi, M.O.; Koike, T.; Mantovani, A.; Meroni, P.L. Role of the MyD88 transduction signaling pathway in endothelial activation by antiphospholipid antibodies. *Blood* **2003**, *101*, 3495–3500. [CrossRef] [PubMed]
27. Espinola, R.G.; Liu, X.; Colden-Stanfield, M.; Hall, J.; Harris, E.N.; Pierangeli, S.S. E-Selectin mediates pathogenic effects of antiphospholipid antibodies. *J. Thromb. Haemost.* **2003**, *1*, 843–848. [CrossRef] [PubMed]
28. Kobayashi, K.; Kishi, M.; Atsumi, T.; Bertolaccini, M.L.; Makino, H.; Sakairi, N.; Yamamoto, I.; Yasuda, T.; Khamashta, M.A.; Hughes, G.R.; et al. Circulating oxidized ldl forms complexes with beta2-glycoprotein I: Implication as an atherogenic autoantigen. *J. Lipid Res.* **2003**, *44*, 716–726. [CrossRef] [PubMed]
29. Li, J.; Kim, K.; Barazia, A.; Tseng, A.; Cho, J. Platelet-neutrophil interactions under thromboinflammatory conditions. *Cell. Mol. Life Sci.* **2015**, *72*, 2627–2643. [CrossRef] [PubMed]
30. Kaul, M.; Erkan, D.; Sammaritano, L.; Lockshin, M.D. Assessment of the 2006 revised antiphospholipid syndrome classification criteria. *Ann. Rheum. Dis.* **2007**, *66*, 927–930. [CrossRef] [PubMed]
31. Gris, J.C.; Bouvier, S. Antiphospholipid syndrome: Looking for a refocusing. *Thromb. Res.* **2013**, *131*, S28–S31. [CrossRef]
32. Artenjak, A.; Lakota, K.; Frank, M.; Cucnik, S.; Rozman, B.; Bozic, B.; Shoenfeld, Y.; Sodin-Semrl, S. Antiphospholipid antibodies as non-traditional risk factors in atherosclerosis based cardiovascular diseases without overt autoimmunity. A critical updated review. *Autoimmun. Rev.* **2012**, *11*, 873–882. [CrossRef] [PubMed]
33. Garcia-Martin, F.; de Arriba, G.; Carrascosa, T.; Moldenhauer, F.; Martin-Escobar, E.; Val, J.; Saiz, F. Anticardiolipin antibodies and lupus anticoagulant in end-stage renal disease. *Nephrol. Dial. Transplant.* **1991**, *6*, 543–547. [CrossRef] [PubMed]
34. Sinico, R.A.; Cavazzana, I.; Nuzzo, M.; Vianelli, M.; Napodano, P.; Scaini, P.; Tincani, A. Renal involvement in primary antiphospholipid syndrome: Retrospective analysis of 160 patients. *Clin. J. Am. Soc. Nephrol.* **2010**, *5*, 1211–1217. [CrossRef] [PubMed]
35. Alchi, B.; Griffiths, M.; Jayne, D. What nephrologists need to know about antiphospholipid syndrome. *Nephrol. Dial. Transplant.* **2010**, *25*, 3147–3154. [CrossRef] [PubMed]
36. Amigo, M.C.; Garcia-Torres, R.; Robles, M.; Bochicchio, T.; Reyes, P.A. Renal involvement in primary antiphospholipid syndrome. *J. Rheumatol.* **1992**, *19*, 1181–1185. [PubMed]
37. Nochy, D.; Daugas, E.; Droz, D.; Beaufils, H.; Grunfeld, J.P.; Piette, J.C.; Bariety, J.; Hill, G. The intrarenal vascular lesions associated with primary antiphospholipid syndrome. *J. Am. Soc. Nephrol.* **1999**, *10*, 507–518. [PubMed]

38. Sciascia, S.; Cuadrado, M.J.; Khamashta, M.; Roccatello, D. Renal involvement in antiphospholipid syndrome. *Nat. Rev. Nephrol.* **2014**, *10*, 279–289. [CrossRef] [PubMed]
39. Gaggl, M.; Aigner, C.; Sunder-Plassmann, G.; Schmidt, A. Thrombotische Mikroangiopathien: Relevante Neuigkeiten für den Intensivmediziner. *Med. Klin. Intensivmed. Notfmed.* **2016**, *111*, 434–439. (In German) [CrossRef] [PubMed]
40. Fakhouri, F.; Noel, L.H.; Zuber, J.; Beaufils, H.; Martinez, F.; Lebon, P.; Papo, T.; Chauveau, D.; Bletry, O.; Grunfeld, J.P.; et al. The expanding spectrum of renal diseases associated with antiphospholipid syndrome. *Am. J. Kidney Dis.* **2003**, *41*, 1205–1211. [CrossRef]
41. Morales, J.M.; Martinez-Flores, J.A.; Serrano, M.; Castro, M.J.; Alfaro, F.J.; Garcia, F.; Martinez, M.A.; Andres, A.; Gonzalez, E.; Praga, M.; et al. Association of early kidney allograft failure with preformed IgA antibodies to beta2-glycoprotein I. *J. Am. Soc. Nephrol.* **2015**, *26*, 735–745. [CrossRef] [PubMed]
42. Cacoub, P.; Wechsler, B.; Piette, J.C.; Beaufils, H.; Herreman, G.; Bletry, O.; Godeau, P. Malignant hypertension in antiphospholipid syndrome without overt lupus nephritis. *Clin. Exp. Rheumatol.* **1993**, *11*, 479–485. [PubMed]
43. Sangle, S.R.; D'Cruz, D.P.; Jan, W.; Karim, M.Y.; Khamashta, M.A.; Abbs, I.C.; Hughes, G.R. Renal artery stenosis in the antiphospholipid (Hughes) syndrome and hypertension. *Ann. Rheum. Dis.* **2003**, *62*, 999–1002. [CrossRef] [PubMed]
44. Ben-Ami, D.; Bar-Meir, E.; Shoenfeld, Y. Stenosis in antiphospholipid syndrome: A new finding with clinical implications. *Lupus* **2006**, *15*, 466–472. [CrossRef] [PubMed]
45. Asherson, R.A.; Hughes, G.R. Primary antiphospholipid syndrome. *Am. J. Med.* **1993**, *94*, 345–346. [CrossRef]
46. Lai, N.S.; Lan, J.L. Renal vein thrombosis in chinese patients with systemic lupus erythematosus. *Ann. Rheum. Dis.* **1997**, *56*, 562–564. [CrossRef] [PubMed]
47. Pons-Estel, G.J.; Cervera, R. Renal involvement in antiphospholipid syndrome. *Curr. Rheumatol. Rep.* **2014**, *16*. [CrossRef] [PubMed]
48. Silvarino, R.; Sant, F.; Espinosa, G.; Pons-Estel, G.; Sole, M.; Cervera, R.; Arrizabalaga, P. Nephropathy associated with antiphospholipid antibodies in patients with systemic lupus erythematosus. *Lupus* **2011**, *20*, 721–729. [CrossRef] [PubMed]
49. Zheng, H.; Chen, Y.; Ao, W.; Shen, Y.; Chen, X.W.; Dai, M.; Wang, X.D.; Yan, Y.C.; Yang, C.D. Antiphospholipid antibody profiles in lupus nephritis with glomerular microthrombosis: A prospective study of 124 cases. *Arthritis Res. Ther.* **2009**, *11*. [CrossRef] [PubMed]
50. Merrill, J.T.; Asherson, R.A. Catastrophic antiphospholipid syndrome. *Nat. Clin. Pract. Rheumatol.* **2006**, *2*, 81–89. [CrossRef] [PubMed]
51. Cervera, R.; Asherson, R.A. Catastrophic antiphospholipid (Asherson's) syndrome. *Br. J. Hosp. Med.* **2008**, *69*, 384–387. [CrossRef] [PubMed]
52. Asherson, R.A.; Cervera, R.; Piette, J.C.; Font, J.; Lie, J.T.; Burcoglu, A.; Lim, K.; Munoz-Rodriguez, F.J.; Levy, R.A.; Boue, F.; et al. Catastrophic antiphospholipid syndrome. Clinical and laboratory features of 50 patients. *Medicine* **1998**, *77*, 195–207. [CrossRef] [PubMed]
53. Cerveny, K.C.; Sawitzke, A.D. Relapsing catastrophic antiphospholipid antibody syndrome: A mimic for thrombotic thrombocytopenic purpura? *Lupus* **1999**, *8*, 477–481. [CrossRef] [PubMed]
54. Cervera, R.; Bucciarelli, S.; Plasin, M.A.; Gomez-Puerta, J.A.; Plaza, J.; Pons-Estel, G.; Shoenfeld, Y.; Ingelmo, M.; Espinos, G. Catastrophic antiphospholipid syndrome (CAPS): Descriptive analysis of a series of 280 patients from the "CAPS registry". *J. Autoimmun.* **2009**, *32*, 240–245. [CrossRef] [PubMed]
55. Asherson, R.A. The catastrophic antiphospholipid syndrome, 1998. A review of the clinical features, possible pathogenesis and treatment. *Lupus* **1998**, *7*, S55–S62. [CrossRef] [PubMed]
56. Brunet, P.; Aillaud, M.F.; San Marco, M.; Philip-Joet, C.; Dussol, B.; Bernard, D.; Juhan-Vague, I.; Berland, Y. Antiphospholipids in hemodialysis patients: Relationship between lupus anticoagulant and thrombosis. *Kidney Int.* **1995**, *48*, 794–800. [CrossRef] [PubMed]
57. Sarnak, M.J. Cardiovascular complications in chronic kidney disease. *Am. J. Kidney Dis.* **2003**, *41*, 11–17. [CrossRef]
58. Ozmen, S.; Danis, R.; Akin, D.; Batun, S. Anticardiolipin antibodies in hemodialysis patients with hepatitis c and their role in fistula failure. *Clin. Nephrol.* **2009**, *72*, 193–198. [PubMed]
59. Roozbeh, J.; Serati, A.R.; Malekhoseini, S.A. Arteriovenous fistula thrombosis in patients on regular hemodialysis: A report of 171 patients. *Arch. Iran. Med.* **2006**, *9*, 26–32. [PubMed]

60. Prieto, L.N.; Suki, W.N. Frequent hemodialysis graft thrombosis: Association with antiphospholipid antibodies. *Am. J. Kidney Dis.* **1994**, *23*, 587–590. [CrossRef]

61. Prakash, R.; Miller, C.C., 3rd; Suki, W.N. Anticardiolipin antibody in patients on maintenance hemodialysis and its association with recurrent arteriovenous graft thrombosis. *Am. J. Kidney Dis.* **1995**, *26*, 347–352. [CrossRef]

62. Quereda, C.; Pardo, A.; Lamas, S.; Orofino, L.; Carcia-Avello, A.; Marcen, R.; Teruel, J.L.; Ortuno, J. Lupus-like in vitro anticoagulant activity in end-stage renal disease. *Nephron* **1988**, *49*, 39–44. [CrossRef] [PubMed]

63. Gronhagen-Riska, C.; Teppo, A.M.; Helantera, A.; Honkanen, E.; Julkunen, H. Raised concentrations of antibodies to cardiolipin in patients receiving dialysis. *BMJ* **1990**, *300*, 1696–1697. [CrossRef] [PubMed]

64. Matsuda, J.; Saitoh, N.; Gohchi, K.; Tsukamoto, M.; Nakamura, K.; Kinoshita, T. Beta 2-glycoprotein i-dependent and-independent anticardiolipin antibody in patients with end-stage renal disease. *Thromb. Res.* **1993**, *72*, 109–117. [CrossRef]

65. Valeri, A.; Joseph, R.; Radhakrishnan, J. A large prospective survey of anti-cardiolipin antibodies in chronic hemodialysis patients. *Clin. Nephrol.* **1999**, *51*, 116–121. [PubMed]

66. Ducloux, D.; Bourrinet, E.; Motte, G.; Chalopin, J.M. Antiphospholipid antibodies as a risk factor for atherosclerotic events in renal transplant recipients. *Kidney Int.* **2003**, *64*, 1065–1070. [CrossRef] [PubMed]

67. Canaud, G.; Bienaime, F.; Noel, L.H.; Royal, V.; Alyanakian, M.A.; Dautzenberg, M.D.; Rabant, M.; Posson, J.; Thervet, E.; Anglicheau, D.; et al. Severe vascular lesions and poor functional outcome in kidney transplant recipients with lupus anticoagulant antibodies. *Am. J. Transplant.* **2010**, *10*, 2051–2060. [CrossRef] [PubMed]

68. Furmanczyk-Zawiska, A.; Baczkowska, T.; Sadowska, A.; Szmidt, J.; Chmura, A.; Durlik, M. Antiphospholipid antibodies in renal allograft recipients. *Transplant. Proc.* **2013**, *45*, 1655–1660. [CrossRef] [PubMed]

69. Vaidya, S.; Wang, C.C.; Gugliuzza, C.; Fish, J.C. Relative risk of post-transplant renal thrombosis in patients with antiphospholipid antibodies. *Clin. Transplant.* **1998**, *12*, 439–444. [PubMed]

70. Wagenknecht, D.R.; Becker, D.G.; LeFor, W.M.; McIntyre, J.A. Antiphospholipid antibodies are a risk factor for early renal allograft failure. *Transplantation* **1999**, *68*, 241–246. [CrossRef] [PubMed]

71. Ducloux, D.; Pellet, E.; Fournier, V.; Rebibou, J.M.; Bresson-Vautrin, C.; Racadot, E.; Fellmann, D.; Chalopin, J.M. Prevalence and clinical significance of antiphospholipid antibodies in renal transplant recipients. *Transplantation* **1999**, *67*, 90–93. [CrossRef] [PubMed]

72. Fernandez-Fresnedo, G.; Lopez-Hoyos, M.; Segundo, D.S.; Crespo, J.; Ruiz, J.C.; de Francisco, A.L.; Arias, M. Clinical significance of antiphospholipid antibodies on allograft and patient outcome after kidney transplantation. *Transplant. Proc.* **2005**, *37*, 3710–3711. [CrossRef] [PubMed]

73. Leskinen, Y.; Lehtimaki, T.; Loimaala, A.; Lautamatti, V.; Kallio, T.; Huhtala, H.; Salenius, J.P.; Saha, H. Carotid atherosclerosis in chronic renal failure-the central role of increased plaque burden. *Atherosclerosis* **2003**, *171*, 295–302. [CrossRef] [PubMed]

74. Kovacic, V.; Ljutic, D.; Dodig, J.; Radic, M.; Duplancic, D. Influence of haemodialysis on early markers of atherosclerosis. *Nephrology* **2008**, *13*, 472–479. [CrossRef] [PubMed]

75. McGregor, D.O.; Buttimore, A.L.; Lynn, K.L.; Yandle, T.; Nicholls, M.G. Effects of long and short hemodialysis on endothelial function: A short-term study. *Kidney Int.* **2003**, *63*, 709–715. [CrossRef] [PubMed]

76. Bertolaccini, M.L.; Amengual, O.; Andreoli, L.; Atsumi, T.; Chighizola, C.B.; Forastiero, R.; de Groot, P.; Lakos, G.; Lambert, M.; Meroni, P.; et al. 14th international congress on antiphospholipid antibodies task force. Report on antiphospholipid syndrome laboratory diagnostics and trends. *Autoimmun. Rev.* **2014**, *13*, 917–930. [CrossRef] [PubMed]

77. Serrano, A.; Garcia, F.; Serrano, M.; Ramirez, E.; Alfaro, F.J.; Lora, D.; de la Camara, A.G.; Paz-Artal, E.; Praga, M.; Morales, J.M. IgA antibodies against beta2-glycoprotein I in hemodialysis patients are an independent risk factor for mortality. *Kidney Int.* **2012**, *81*, 1239–1244. [CrossRef] [PubMed]

78. Malyszko, J.; Malyszko, J.S.; Bachorzewska-Gajewska, H. Cardiovascular risk in chronic renal disease and transplantation prevention and management. *Expert Opin. Pharmacother.* **2005**, *6*, 929–943. [CrossRef] [PubMed]

79. Hadhri, S.; Rejeb, M.B.; Belarbia, A.; Achour, A.; Skouri, H. Hemodialysis duration, human platelet antigen HPA-3 and IgA isotype of anti-beta2glycoprotein I antibodies are associated with native arteriovenous fistula failure in Tunisian hemodialysis patients. *Thromb. Res.* **2013**, *131*, e202–e209. [CrossRef] [PubMed]

80. Serrano, M.; Martinez-Flores, J.A.; Castro, M.J.; Garcia, F.; Lora, D.; Perez, D.; Gonzalez, E.; Paz-Artal, E.; Morales, J.M.; Serrano, A. Renal transplantation dramatically reduces IgA anti-beta-2-glycoprotein I antibodies in patients with endstage renal disease. *J. Immunol. Res.* **2014**, *2014*, 143–146. [CrossRef] [PubMed]

81. Morales, J.M.; Serrano, M.; Martinez-Flores, J.A.; Perez, D.; Castro, M.J.; Sanchez, E.; Garcia, F.; Rodriguez-Antolin, A.; Alonso, M.; Gutierrez, E.; et al. The presence of pretransplant antiphospholipid antibodies IgA anti-beta-2-glycoprotein I as a predictor of graft thrombosis after renal transplantation. *Transplantation* **2016**. [CrossRef] [PubMed]

82. Ruiz-Irastorza, G.; Hunt, B.J.; Khamashta, M.A. A systematic review of secondary thromboprophylaxis in patients with antiphospholipid antibodies. *Arthritis Rheum.* **2007**, *57*, 1487–1495. [CrossRef] [PubMed]

83. Tripodi, A.; de Groot, P.G.; Pengo, V. Antiphospholipid syndrome: Laboratory detection, mechanisms of action and treatment. *J. Intern. Med.* **2011**, *270*, 110–122. [CrossRef] [PubMed]

84. Urbanus, R.T.; Siegerink, B.; Roest, M.; Rosendaal, F.R.; de Groot, P.G.; Algra, A. Antiphospholipid antibodies and risk of myocardial infarction and ischaemic stroke in young women in the ratio study: A case-control study. *Lancet Neurol.* **2009**, *8*, 998–1005. [CrossRef]

85. Erkan, D.; Yazici, Y.; Peterson, M.G.; Sammaritano, L.; Lockshin, M.D. A cross-sectional study of clinical thrombotic risk factors and preventive treatments in antiphospholipid syndrome. *Rheumatology (Oxford)* **2002**, *41*, 924–929. [CrossRef] [PubMed]

86. Tarr, T.; Lakos, G.; Bhattoa, H.P.; Shoenfeld, Y.; Szegedi, G.; Kiss, E. Analysis of risk factors for the development of thrombotic complications in antiphospholipid antibody positive lupus patients. *Lupus* **2007**, *16*, 39–45. [CrossRef] [PubMed]

87. Hereng, T.; Lambert, M.; Hachulla, E.; Samor, M.; Dubucquoi, S.; Caron, C.; Launay, D.; Morell-Dubois, S.; Queyrel, V.; Hatron, P.Y. Influence of aspirin on the clinical outcomes of 103 anti-phospholipid antibodies-positive patients. *Lupus* **2008**, *17*, 11–15. [CrossRef] [PubMed]

88. Erkan, D.; Merrill, J.T.; Yazici, Y.; Sammaritano, L.; Buyon, J.P.; Lockshin, M.D. High thrombosis rate after fetal loss in antiphospholipid syndrome: Effective prophylaxis with aspirin. *Arthritis Rheum.* **2001**, *44*, 1466–1467. [CrossRef]

89. Marson, P.; Bagatella, P.; Bortolati, M.; Tison, T.; de Silvestro, G.; Fabris, F.; Pengo, V.; Ruffatti, A. Plasma exchange for the management of the catastrophic antiphospholipid syndrome: Importance of the type of fluid replacement. *J. Intern. Med.* **2008**, *264*, 201–203. [CrossRef] [PubMed]

90. Erkan, D.; Lockshin, M.D. New approaches for managing antiphospholipid syndrome. *Nat. Clin. Pract. Rheumatol.* **2009**, *5*, 160–170. [CrossRef] [PubMed]

91. Lonze, B.E.; Zachary, A.A.; Magro, C.M.; Desai, N.M.; Orandi, B.J.; Dagher, N.N.; Singer, A.L.; Carter-Monroe, N.; Nazarian, S.M.; Segev, D.L.; et al. Eculizumab prevents recurrent antiphospholipid antibody syndrome and enables successful renal transplantation. *Am. J. Transplant.* **2014**, *14*, 459–465. [CrossRef] [PubMed]

92. Connolly, S.J.; Ezekowitz, M.D.; Yusuf, S.; Eikelboom, J.; Oldgren, J.; Parekh, A.; Pogue, J.; Reilly, P.A.; Themeles, E.; Varrone, J.; et al. Dabigatran versus warfarin in patients with atrial fibrillation. *N. Engl. J. Med.* **2009**, *361*, 1139–1151. [CrossRef] [PubMed]

93. Cohen, H.; Machin, S.J. Antithrombotic treatment failures in antiphospholipid syndrome: The new anticoagulants? *Lupus* **2010**, *19*, 486–491. [CrossRef] [PubMed]

94. Kolyada, A.; de Biasio, A.; Beglova, N. Identification of the binding site for fondaparinux on beta2-glycoprotein I. *Biochim. Biophys. Acta* **2013**, *1834*, 2080–2088. [CrossRef] [PubMed]

95. Ruiz-Irastorza, G.; Crowther, M.; Branch, W.; Khamashta, M.A. Antiphospholipid syndrome. *Lancet* **2010**, *376*, 1498–1509. [CrossRef]

96. Cervera, R.; Serrano, R.; Pons-Estel, G.J.; Ceberio-Hualde, L.; Shoenfeld, Y.; de Ramon, E.; Buonaiuto, V.; Jacobsen, S.; Zeher, M.M.; Tarr, T.; et al. Morbidity and mortality in the antiphospholipid syndrome during a 10-year period: A multicentre prospective study of 1000 patients. *Ann. Rheum. Dis.* **2015**, *74*, 1011–1018. [CrossRef] [PubMed]

97. Martinez-Flores, J.A.; Serrano, M.; Perez, D.; Lora, D.; Paz-Artal, E.; Morales, J.M.; Serrano, A. Detection of circulating immune complexes of human IgA and beta 2 glycoprotein I in patients with antiphospholipid syndrome symptomatology. *J. Immunol. Methods* **2015**, *422*, 51–58. [CrossRef] [PubMed]

98. Martinez-Flores, J.A.; Serrano, M.; Perez, D.; Camara, A.G.; Lora, D.; Morillas, L.; Ayala, R.; Paz-Artal, E.; Morales, J.M.; Serrano, A. Circulating immune complexes of IgA bound to beta2 glycoprotein are strongly associated with the occurrence of acute thrombotic events. *J. Atheroscler. Thromb.* **2016**. [CrossRef] [PubMed]

99. Sciascia, S.; Sanna, G.; Murru, V.; Roccatello, D.; Khamashta, M.A.; Bertolaccini, M.L. Anti-prothrombin (aPT) and anti-phosphatidylserine/prothrombin (aPS/PT) antibodies and the risk of thrombosis in the antiphospholipid syndrome. A systematic review. *Thromb. Haemost.* **2014**, *111*, 354–364. [CrossRef] [PubMed]

100. Iaccarino, L.; Ghirardello, A.; Canova, M.; Zen, M.; Bettio, S.; Nalotto, L.; Punzi, L.; Doria, A. Anti-annexins autoantibodies: Their role as biomarkers of autoimmune diseases. *Autoimmun. Rev.* **2011**, *10*, 553–558. [CrossRef] [PubMed]

101. Canaud, G.; Bienaime, F.; Tabarin, F.; Bataillon, G.; Seilhean, D.; Noel, L.H.; Dragon-Durey, M.A.; Snanoudj, R.; Friedlander, G.; Halbwachs-Mecarelli, L.; et al. Inhibition of the mtorc pathway in the antiphospholipid syndrome. *N. Engl. J. Med.* **2014**, *371*, 303–312. [CrossRef] [PubMed]

antibodies

MDPI

Review

Antiphospholipid Antibodies: From General Concepts to Its Relation with Malignancies

José A. Gómez-Puerta [1,2], Gerard Espinosa [3] and Ricard Cervera [3,*]

1   Grupo de Inmunología Celular e Inmunogenética y Grupo de Reumatología, Universidad de Antioquia, Medellín 05004, Antioquia, Colombia
2   Consultor de Reumatología, Dinámica IPS, Medellín 050015, Antioquia, Colombia
3   Department of Autoimmune Diseases, Hospital Clínic, Villarroel, 170, Barcelona 08036, Catalonia, Spain
*   Correspondence: rcervera@clinic.cat; Tel.: +34-93-227-5774; Fax: +34-93-227-1707

Academic Editor: Dimiter S. Dimitrov
Received: 10 May 2016; Accepted: 5 July 2016; Published: 2 August 2016

Abstract: Antiphospholipid syndrome (APS) is an adquired autoimmune pro-thrombotic disease characterized by arterial and/or venous thrombosis and/or fetal losses associated with the persistent presence of antiphospholipid antibodies (aPL) detectable by solid phase assays (anticardiolipin (aCL) and anti-$\beta 2$ glycoprotein I, $\beta 2$GPI) and/or functional coagulation test (lupus anticoagulant (LA)). Most patients with typical APS manifestations have the presence of one or more of conventional aPL, but, some patients might exhibit clinical features related with APS but with persistent negative determinations of "classic" aPL (seronegative APS). Expanding the network of autoantibodies in patients highly suspected of having APS but who have normal results from a conventional test using new antibodies (i.e., phosphatidylserine/prothrombin and $\beta 2$GPI domain 1) would increase the diagnosis. Thrombosis is one of the leading causes of death among patients with cancer, representing up to 15% of all deaths. Cancer increases the risk of thrombosis and chemotherapy is further associated with a higher risk of thrombosis. In addition, aPL may contribute to an increased risk of thrombosis in patients with malignancies, although the levels do not seem to reflect their pathogenicity. Several malignancies, particularly hematological and lymphoproliferative malignancies, may indeed be associated with the generation of aPL but do not necessarily enhance the thrombophilic risk in these patients.

Keywords: antiphospholipid antibodies; lupus anticoagulant; anticardiolipin antibodies; cancer; malignancies; catastrophic antiphospholipid syndrome

1. Introduction

Antiphospholipid syndrome (APS) is defined by the presence of arterial and venous thromboses and pregnancy morbidity (miscarriages, fetal deaths, premature births), in the company of antiphospholipid antibodies (aPL); namely, lupus anticoagulant (LA), anticardiolipin antibodies (aCL), or anti-$\beta 2$ glycoprotein-I (anti-$\beta 2$GPI) antibodies. APS can occur in patients having neither clinical nor laboratory evidence of another definable condition (primary APS), or it may be associated with other diseases, mainly systemic lupus erythematosus (SLE), and occasionally with other autoimmune conditions (Sjögren syndrome, systemic vasculitis, rheumatoid arthritis, among others), infections, drugs, and malignancies [1].

Despite the prevalence of aPL in general population is up to 5%, only a small proportion of patients develop APS. Some epidemiological studies estimates that the incidence of APS is around 5 new cases per 100,000 persons per year and the prevalence around 40–50 cases per 100,000 persons [2]. The prevalence increases in the elderly and in those with chronic disease. Very recently, a population-based cohort was conducted in Germany (Gutenberg Health

Study) including 5000 subjects (2540 men, 2460 women) from April 2007 to October 2008 [3]. aCL, anti-β2GPI and anti-β2GPI domain 1 were measured in 4977 subjects. The authors found a strong age-dependent increase of both aCL and anti-β2GPI IgM, while aPL IgG titers were stable or tended to decrease with age [3].

In 2011, an international group, APS ACTION (AntiPhospholipid Syndrome Alliance For Clinical Trials and InternatiOnal Networking), gathered with the aim of planning several studies on aPL-related syndromes. Its primary mission involves the prevention, treatment, and cure of aPL-associated clinical manifestations through high-quality, multicentre, and multidisciplinary clinical research. Recently, the APS ACTION group published a literature review focused on the prevalence of aPL in the general population for 4 different outcomes of APS: stroke, myocardial infarction (MI), deep vein thrombosis (DVT) and pregnancy morbidity. APS action group estimated that aPL are positive in approximately 13% of patients with stroke, 11% with MI, 9.5% of patients with DVT and 6% of patients with pregnancy morbidity [4].

One of the pivotal studies on APS was conducted by The Euro-Phospholipid Group. This International group of experts from 13 European countries analyzed the prevalence of the most relevant clinical and immunological features in a cohort of 1,000 APS patients. Stroke and transient ischemic attacks were the most common arterial manifestations (19.8%, 11.1% respectively), followed by leg ulcers (5.5%), MI (5.5%), and amaurosis fugax (5.4%). Regarding venous events, the most frequent features were: DVT (38.9%), pulmonary embolism (14.1%), and superficial thrombophlebitis (11.7%). Other clinical manifestations included thrombocytopenia (29.6%), *livedo reticularis* (24.1%), heart valve lesions (14.3%), hemolytic anemia (9.7%) and epilepsy (7%) among others [5].

One uncommon, but often lethal variant of APS, characterized by a rapid and progressive thrombosis (mainly small vessel trombosis) is known as catastrophic APS (CAPS) [6]. Fortunately the prevalence of the catastrophic APS is rare (<1% of all cases of APS) but its potentially fatal outcome emphasizes its significance in clinical practice. In order to summarize all the published case reports as well as the new diagnosed cases from all over the world, an international registry of patients with catastrophic APS ("CAPS Registry") was created in 2000 by the European Forum on Antiphospholipid Antibodies. Currently, CAPS registry includes clinical, laboratory and therapeutic data of around 500 cases. This registry can be freely checked on the Internet https://ontocrf.grupocostaisa.com/es/web/caps/home.

2. aPL Antibodies

aPL antibodies are a heterogenous group of autoantibodies directed against anionic phospholipids or protein-phospholipid complexes, measured in solid phase immunoassays such as aCL or as an activity (functional assays) which prolongs phospholipid-dependent coagulation assays, the so-called LA. There are three well described and validated aPL antibodies included in the current revised classification criteria ("Sydney criteria") [7], including aCL (IgG and IgM), LA and β2GPI (IgG and IgM). Accoding to revised classification criteria, aCL antibodies are considered positive when they are present in serum or plasma at medium or high titers on two or more occasions, at least 12 weeks apart, measured by a standardized ELISA technique. LA is positive when it is present in plasma, on two or more occasions at least 12 weeks apart, detected according to the guidelines of the International Society on Thrombosis and Hemostasis. Finally, Anti-β2GP1 antibodies are considered positive in serum or plasma, on two or more occasions, at least 12 weeks apart, measured by a standardized ELISA technique. Some other autoantibodies directed against anionic phospholipids or serological assays that detect antibodies to coagulation proteins have been reported during the last years [8]. However, not all of them have been replicated in other groups or have been standardized using conventional techniques (Table 1).

APS patients nor only have the presence of aPL antibodies, but also a wide variety of autoantibodies in secondary APS patients, including antinuclear antibodies, anti-dsDNA antibodies and extractable nuclear antigen antibodies, among others. The most common immunological features in APS patients are collected in Table 2.

Table 1. Criteria and non-criteria antiphospholipid antibodies (aPL) antibodies.

Criteria aPL
aCL IgG and IgM
Anti-β2GPI IgG and IgM
LA

Other non-criteria aPL
aCL IgA
Anti-β2GPI IgA
Anti-annexin A2
Anti vimentin/cardiolipin complex
Anti-annexin A5
Antiphosphatidylethanolamine
Antiphosphatidylinositol
Anti PT/PS *
Anti-β2GPI Domain I *

* New promising criteria aPL.

Table 2. Most common autoantibodies in the antiphospholipid syndrome (APS), according to the "Euro-Phospholipid Project" (including patients with Primary APS and associated APS, mainly systemic lupus erythematosus (SLE)) [5].

Autoantibody	%
aCL	87.9
IgG and IgM aCL	32.1
IgG aCL alone	43.6
IgM aCL alone	12.2
LA	53.6
LA alone	12.1
LA and aCL	41.5
ANA	59.7
Anti-dsDNA	29.2
Anti-Ro/SS-A	14
Anti-La/SS-B	5.7
Anti-RNP	5.9
Anti-Sm	5.5
Rheumatoid factor	7.8

No information about β2GPI antibodies was available.

3. Seronegative APS

Seronegative APS is defined as patients with typical manifestations suggestive of APS (i.e., *livedo reticularis*, recurrent pregnancy losses, DVT or thrombocytopenia) but who have tested persistently negative for conventional aPL on several occasions. The term seronegative APS was quoted for the first time by Hughes and Khamashta [9]. Some potential explanations for seronegative APS include (1) antibody consumption during an acute thrombotic episode; (2) transient negativity of previously positive aPL patients (unlikely); and (3) a more realistic one: antibodies to the heterogeneous aPL family against protein and protein-bound phospholipids which have not been identified to date. The most promising of "non-classic" aPL are antibodies to phospholipid-protein complexes (vimentin/cardiolipin complex), antibodies against phospholipid-binding plasma proteins (prothrombin (PT), protein C, protein S, annexin V, and domains of β2GPI) [10]; phospholipid-protein complexes (vimentin/cardiolipin complex); and anionic phospholipids other than cardiolipin (phosphatidylserine (PS), phosphatidylinositol) [11] and antibodies to the complex PS/PT [12].

Expanding the network of autoantibodies in patients with normal results from a classic test (aCL, LA and/or β2GPI) using new antibodies (i.e., PT/PS and β2GPI domain 1) in patients with suspected APS would increase the diagnostic capability of detection of new cases of APS formerly labeled as "seronegative" cases [13]. Recent reports have shown that anti β2GPI domain 1 antibodies might achieve a specificity as high as 99.5% for patients with APS and thrombosis events [14].

4. aPL and Malignancies

It is known as Trosseau's syndrome the association between neoplastic disease and a thromboembolic disorder made by Armand Trousseau in 1865 [15]. During decades, the relationship between thrombosis and cancer was well documented. During the last 40 years, several case reports of aCL in patients with thrombotic events and malignant conditions, including hematological, lymphoproliferative disorders and solid tumors have been published. The increasing knowledge of aPL in the pathogenesis of vascular occlusions established a close link between aPL and malignancies.

There has been experimental work demonstrating tumor growth with agents activating blood coagulation and regression with coagulation inhibitors. Fibrin generation has also been associated with accelerated tumor growth and tumor cells themselves may be responsible for the production of compounds resulting in this mechanism of thrombosis [16].

Tumoral cells activate coagulation system through different pathways, interacting with clot cells, platelets and fibrinolytic systems to generate thrombin. In addition, some other endothelial factors, such as fibrin and tissue factor might play a role in the clotting formation mediated via fibrin deposition and platelet activation [17].

Several mechanisms have been suggested to explain the association between aPL and cancer including the following: (1) production of autoantibodies as a response to tumor antigens; (2) secretion of aCL from tumor cells; and (3) production of monoclonal immunoglobulins with LA and aCL activities [18].

In addition, some other clinical factors contribute to the risk of thrombosis including immobilization or intravenous catheters. Furthermore, the risk of thrombosis appears to be highest during the initial hospitalization and onset of chemotherapy, as well as at the time of disease progression [19].

Information about the association between aPL and malignancies (solid tumors and hematological malignancies) are heterogeneous and some, but not all, included information regarding clinical features. Different series of patients with aPL and malignancies are summarized on Table 3.

One of the main studies in the field, was conducted in early 90's in Montepellier, France [20]. The study included 1014 patients who were tested at entry for aCL, been carcinoma was the most frequently associated disease. Only 7.1% of subjects were positive for aPL. Among them, 14 had a history of carcinoma, 9 had active malignant and 5 were in remission. The main related malignancies found were prostatic adenocarcinoma, breast carcinoma, ovarian carcinoma, and colon adenocarcinoma.

Table 3. Series of patients with Malignancies and aPL.

Author (year) REF	No. Patients	Mean Age (years)	Female Gender (%)	Solid Tumors	Haematological Neoplasms	aPL	APS	Thrombotic Manifestations
Gómez-Puerta et al. (2006) [18]	120	56	48	Renal cell carcinoma 6% Primary unknown origin 5% Lung adenocarcinoma 5% Breast cancer 5%	B Cell lymphoma 8% Spleen lymphoma 7% Chronic myeloid leukaemia 5%	LA 67% aCL (67%) (54 aCL IgG and 20 aCL IgM, B2GP1 6%)	Primary APS in 22 (18%) patients *	Thrombotic manifestations 71% APS Sapporo criteria in 21%
Zuckerman et al. (1995) [21]	216	67	45	Colorectal cancer 17% Lung carcinoma 12% Breast cancer 9%	NH lymphoma 10% Multiple myeloma 5%	47 (22%) were aCL positive, compared with 3 in the control group	NA	Thromboembolic events in 13 patients
Miesbach et al. (2006) [22]	58	59	55	67% solid tumors	NH lymphoma 15% Myeloproliferative diseases 8% Acute leukaemia 3%	LA 48% IgG aCL 40% IgM aCL 62%	3 patients APS associated to SLE *	In patients with solid tumours 46% and 32% in haematological
Font et al. (2011) [23]	258	58	43	Colorectal 24% Lung cancer 14% Breast cancer 14% Urinary tract tumors 14%	–	aPL positive 8%	4 patients met APS criteria **	Four patients met classification criteria for APS
Yooon et al. (2003) [24]	33	58	57	Non-Small cell lung cancer 27% Colorectal 15% Ovarian 12%	NH lymphoma 6%	aPL 60% IgA B2GP1 46% aCL IgG 6.7% aCL IgM 16.7%	NA	Venous thrombosis 87% Arterial thrombosis 24%
de Meis et al. (2009) [25]	105	NR	NR	Lung carcinoma 100%	–	In thrombosis group LA 36%, B2GP1 9%	NA	Presence of β2GP1 IgM was negatively correlated with thrombosis
Bazzan et al. (2009) [26]	137	61	73	Breast cancer 56% Colorectal 16% Head-neck 12%	Haematological disease 11%	Overall aPL 24%, LA 5.8%, aCL IgG 8.8%, aCL IgM 3.6% B2GP1 IgG 3.6%, B2GP1 IgM 2.2%	5 patients met APS criteria **	Nine (6.5%) patients with VTE
Vassalo et al. (2014) [27]	95	63	44	Solid tumors 79% Gastrointestinal 23% Head and neck 12% Brain 11%	Hematological 21% NH lymphoma 10% Hodgkin's lymphoma 4%, Acute leukaemia 4%	LA 61% IgA B2GP1 31% aCL 1%	NA	Venous thrombosis in 4% All patients admitted at ICU

* Sapporo APS Criteria; ** Sydney APS Criteria.

Another study evaluated the prevalence of aPL in 216 consecutive patients admitted with a biopsy/cytology proven neoplastic disease [21]. Additionally, the study included as a control group 88 age-matched healthy subjects. aPL were more prevalent among cancer patients (22%) in comparison with control group (3%). Among cancer patients, thrombotic rates were higher in aCL-positive patients (13 out of 47, 28%) than in aCL-negative patients (24/169, 14%).

A German group, retrospectively studied the presence of aPL and thrombotic manifestations in a cohort of 58 patients with previous history of neoplasia (39 solid tumours and 19 hematologic/lymphoproliferative disease) [22]. LA was positive in 46% of patients, IgG aCL in 41%, IgM aCL in 64%, and 55% of patients had elevated levels of both. Of the patients with solid tumors, 18/39 (46%) patients had thromboembolic features of APS. Of the patients with hematologic and lymphoproliferative malignancies, only 6/19 (32%) suffered from thromboembolic events.

Some years ago, we performed a literature review of cases with aPL related with solid tumors, lymphoproliferative and hematological malignancies [18]. Given the heterogeneity of information, a wide list of neoplastic disorders were identified. B-cell lymphoma, splenic lymphoma, chronic myeloid leukemia, and non-Hodgkin's lymphoma were the most common hematologic disorders. Regarding solid tumors, renal cell carcinoma, primary tumor with unknown origin, lung adenocarcinoma and breast carcinoma were the main solid tumors related with aPL. In that series, 29 out of 120 cases of malignancy were diagnosed after the thrombotic manifestation of APS and in 41 cases, the diagnosis of both conditions (APS and cancer) was made at the same time.

A big cohort of cytology/histologically-confirmed solid tumor patients with an active disease with a new diagnosis of venous thromboembolism (VTE) was recently published [23]. In addition, two age-sex matched groups were included (one group of outpatients with diagnosis of solid tumors and no history of VTE and one group of healthy individuals without previous thrombotic events, history of abortions or autoimmune disease). Finally, the study included 258 patients with cancer and VTE, 142 patients with cancer without VTE and 258 healthy controls. aPL antibodies (aCL, LA and B2GPI) were measured in the first 72 hours after VTE and at least 12 weeks after the first in aPL positive patients and healthy controls. aPL were more prevalent in patients with cancer and VTE (21 out of 258; 8.1%) compared to cancer patients VTE negative (2 out of 142; 1.4%) and healthy subjects (2 out of 258; 0.8%). LA and aCL IgG were the most-frequent aPL, followed by β2GPI IgG antibodies. The authors concluded that in comparison with cancer patients without VTE and healthy individuals, cancer patients with VTE had an elevated prevalence of aPL. In addition, it was suggested that the presence of aPL may identify a subset of cancer patients who are at high risk of developing thrombotic complications. For instance, those patients with persistent aPL positivity, especially those with triple positivity (aCL, LA and β2GPI) need a closer monitoring due to a higher risk of new episodes of thrombosis, even in patients under anticoagulation treatment.

Since the former cited studies place a focus on "classic" antibodies included on Sydney criteria, it is possible that some other cases of malignancy-associated thrombosis might be attributed to "novel" aPL. Whether these subgroups of patients represent a different clinical outcome or a worse prognosis requires further analysis.

5. Catastrophic APS

As we cited previously, Catastrophic APS is an uncommon form of presentation of APS, in the majority of case characterized by severe thrombotic complications predominantly affecting small vessels of organs, however, some patients can developed large vessel involvement as occurs in classic forms of APS [7]. Of the 488 cases included so far in the CAPS registry (https://ontocrf.grupocostaisa.com/es/web/caps/home) 50 (10.2%) patients had malignancies; 28 (56%) were female and 22 (44%) were male. The mean age was 46.9 years (SD 22 years). Of the patients, 3 (6%) had SLE, and 44 (88%) had a primary APS. LA were detected in 40 patients (80%), IgG aCL in 35 patients (70%), and IgM aCL in 23 patients (46%). Almost half of patients had thrombocytopenia (42% of cases).

Combined therapy with anticoagulation, steroids, plasma exchange and/or intravenous immunoglobulins is the standard treatment of patients with CAPS. Despite that treatment, survival rate of patients with CAPS still poor. The outcome of patients with CAPS is worse in the presence of an additional malignancy than when no malignancy is present. Only 38% of CAPS patients with malignancies recovered in comparison with 64% of patients without malignancies ($p < 0.05$). Treatment modalities, however, did not differ significantly between these patients. Only around 40% of CAPS patients with malignancies improved. This may be due to the additional presence of the malignancy and to the older age of the patients. Other potential confounding factors such as concomitant chemotherapy treatment, cancer stage, disease duration, comorbidities, among others were not assessed.

6. Conclusions

aPL are a heterogenous group of autoantibodies directed at phospholipid binding proteins. The "classic" aPL include LA, aCL and anti-β2GP1 antibodies; however, some autoantibodies directed against anionic phospholipids or coagulation proteins have been reported during the last years. The presence of aPL may contribute to an increased risk of thrombosis in patients with malignancies, although the levels do not seem to reflect their pathogenicity. Probably, the persistence of aPL over time, and the combination of two or more aPL, has more specific weight on the risk of thrombosis during follow-up.

According to previous data, it is desirable to perform a complete search of aPL in those patients with cancer and thrombosis. Conversely, it also important, rule out a neoplastic disorder in those patients with APS (either classic forms or CAPS) with a new episode of thrombosis despite an adequate range of anticoagulation.

In the future, it will be desirable to identify different profiles of patients with APS, according their thrombotic and non-thrombotic features and according their aPL panel. Distinguishing the different profiles of patients will, no doubt, present an opportunity to treat in a better way patients suffering from APS, including those patients with malignances related with aPL.

Acknowledgments: José A. Gómez-Puerta was supported by Colciencias (convocatoria 656 de 2014).

Author Contributions: José A. Gómez-Puerta, Gerard Espinosa and Ricard Cervera wrote the review.

Conflicts of Interest: The authors declare no conflict of interest.

Abbreviations

The following abbreviations are used in this manuscript:

aCL	Anticardiolipin
Anti-β2GP1	Anti-beta 2 glycoprotein 1
aPL	Anti-phospholipid antibodies
APS	Antiphospholipid syndrome
CAPS	Catastrophic Antiphospholipid Syndrome
DVT	Deep vein thrombosis
LA	Lupus anticoagulant
MI	Myocardial infarction
PS	phosphatidylserine
PT	Prothrombin
SLE	Systemic lupus erythematosus

References

1. Gómez-Puerta, J.A.; Cervera, R. Diagnosis and classification of the antiphospholipid syndrome. *J. Autoimmun.* **2014**, *48–49*, 20–25. [CrossRef] [PubMed]
2. Biggioggero, M.; Meroni, P.L. The geoepidemiology of the antiphospholipid antibody syndrome. *Autoimmun. Rev.* **2010**, *9*, A299–A304. [CrossRef] [PubMed]

3. Manukyan, D.; Rossmann, H.; Schulz, A.; Zeller, T.; Pfeiffer, N.; Binder, H.; Münzel, T.; Beutel, M.E.; Müller-Calleja, N.; Wild, P.S.; et al. Distribution of antiphospholipid antibodies in a large population-based German cohort. *Clin. Chem. Lab. Med.* **2016**. [CrossRef]
4. Andreoli, L.; Chighizola, C.B.; Banzato, A.; Pons-Estel, G.J.; de Jesus, G.R.; Erkan, D. The estimated frequency of antiphospholipid antibodies in patients with pregnancy morbidity, stroke, myocardial infarction, and deep vein thrombosis. *Arthritis Care Res. (Hoboken)* **2013**, *65*, 1869–1873. [CrossRef] [PubMed]
5. Cervera, R.; Khamashta, M.A.; Shoenfeld, Y.; Camps, M.T.; Jacobsen, S.; Kiss, E.; Zeher, M.M.; Tincani, A.; Kontopoulou-Griva, I.; Galeazzi, M.; et al. Euro-Phospholipid Project Group European Forum on Antiphospholipid Antibodies. Morbidity and mortality in the antiphospholipid syndrome during a 5-year period: A multicentre prospective study of 1000 patients. *Arthritis Rheum.* **2002**, *46*, 1019–1027. [CrossRef] [PubMed]
6. Cervera, R. Lessons from the "Euro-phospholipid" project. *Autoimmun. Rev.* **2008**, *7*, 174–178. [CrossRef] [PubMed]
7. Asherson, R.A. The catastrophic antiphospholipid syndrome. *J. Rheumatol.* **1992**, *19*, 508–512. [PubMed]
8. Miyakis, S.; Lockshin, M.; Atsumi, T.; Branch, D.; Brey, R.; Cervera, R.; Derksen, R.; de Groot, P.; Koike, T.; Meroni, P. International consensus statement on an update of the classification criteria for definite antiphospholipid syndrome (APS). *J. Thromb. Haemost.* **2006**, *4*, 295–306. [CrossRef] [PubMed]
9. Ardila-Suarez, O.; Gómez-Puerta, J.A.; Khamashta, M.A. Laboratory and diagnosis of antiphospholipid syndrome: From an historical perspective to the emergence of new autoantibodies. *Med. Clin. (Barc.)* **2016**, *146*, 555–560. [CrossRef] [PubMed]
10. Hughes, G.R.; Khamashta, M.A. Seronegative antiphospholipid syndrome. *Ann. Rheum. Dis.* **2003**, *62*. [CrossRef]
11. Cousins, L.; Pericleous, C.; Khamashta, M.; Bertolaccini, M.L.; Ioannou, Y.; Giles, I.; Rahman, A. Antibodies to domain I of β-2-glycoprotein I and IgA antiphospholipid antibodies in patients with 'seronegative' antiphospholipid syndrome. *Ann. Rheum. Dis.* **2015**, *74*, 317–319. [CrossRef] [PubMed]
12. Nayfe, R.; Uthman, I.; Aoun, J.; Saad Aldin, E.; Merashli, M.; Khamashta, M.A. Seronegative antiphospholipid syndrome. *Rheumatology (Oxf.)* **2013**, *52*, 1358–1367. [CrossRef] [PubMed]
13. Sciascia, S.; Khamashta, M.A.; Bertolaccini, M.L. New tests to detect antiphospholipid antibodies: Antiprothrombin (aPT) and anti-phosphatidylserine/prothrombin (aPS/PT) antibodies. *Curr. Rheumatol. Rep.* **2014**, *16*, 415–425. [CrossRef] [PubMed]
14. Mahler, M.; Albesa, R.; Zohoury, N.; Bertolaccini, M.L.; Ateka-Barrutia, O.; Rodriguez-Garcia, J.L.; Norman, G.L.; Khamashta, M.A. Autoantibodies to domain 1 of beta 2 glycoprotein I determined using a novel chemiluminescence immunoassay demonstrate association with thrombosis in patients with antiphospholipid syndrome. *Lupus* **2016**, *25*, 911–916. [CrossRef] [PubMed]
15. Trousseau, A. Phlegmasia alba dolens. In *Clinique Medical de L'Hotel Dieu de Paris*; New Sydenham Society: London, UK, 1865; Volume 3, p. 94.
16. Asherson, R.A. Antiphospholipid antibodies, malignancies and paraproteinemias. *J. Autoimmun.* **2000**, *15*, 117–122. [CrossRef] [PubMed]
17. Rickles, F.R.; Patierno, S.; Fernandez, P.M. Tissue factor, thrombin, and cancer. *Chest* **2003**, *124*, 58S–68S. [CrossRef] [PubMed]
18. Gómez-Puerta, J.A.; Cervera, R.; Espinosa, G.; Aguilo, S.; Bucciarelli, S.; Ramos-Casals, M.; Ingelmo, M.; Asherson, R.A.; Font, J. Antiphospholipid antibodies associated with malignancies: Clinical and pathological characteristics of 120 patients. *Semin. Arthritis Rheum.* **2006**, *35*, 322–332. [CrossRef] [PubMed]
19. Lyman, G.H. Venous thromboembolism in the patient with cancer: Focus on burden of disease and benefits of thromboprophylaxis. *Cancer* **2011**, *117*, 1334–1339. [CrossRef] [PubMed]
20. Schved, J.F.; Dupuy-Fons, C.; Biron, C.; Quere, I.; Janbon, C. A prospective epidemiological study on the occurrence of antiphospholipid antibody: The Montpellier Antiphospholipid (MAP) Study. *Haemostasis* **1994**, *24*, 175–182. [CrossRef] [PubMed]
21. Zuckerman, E.; Toubi, E.; Golan, T.D.; Rosenvald-Zuckerman, T.; Shmuel, Z.; Yeshurun, D. Increased thromboembolic incidence in anti-cardiolipin-positive patients with malignancy. *Br. J. Cancer* **1995**, *72*, 447–451. [CrossRef] [PubMed]
22. Miesbach, W.; Scharrer, I.; Asherson, R. Thrombotic manifestations of the antiphospholipid syndrome in patients with malignancies. *Clin. Rheumatol.* **2006**, *25*, 840–844. [CrossRef] [PubMed]

23. Font, C.; Vidal, L.; Espinosa, G.; Tàssies, D.; Monteagudo, J.; Farrús, B.; Visa, L.; Cervera, R.; Gascon, P.; Reverter, J.C. Solid cancer, antiphospholipid antibodies, and venous thromboembolism. *Autoimmun. Rev.* **2011**, *10*, 222–227. [CrossRef] [PubMed]
24. Yoon, K.H.; Wong, A.; Shakespeare, T.; Sivalingam, P. High prevalence of antiphospholipid antibodies in Asian cancer patients with thrombosis. *Lupus* **2003**, *12*, 112–116. [CrossRef] [PubMed]
25. de Meis, E.; Pinheiro, V.R.; Zamboni, M.M.; Guedes, M.T.; Castilho, I.A.; Martinez, M.M. Clotting, immune system, and venous thrombosis in lung adenocarcinoma patients: A prospective study. *Cancer Investig.* **2009**, *27*, 989–997.
26. Bazzan, M.; Montarulli, B.; Vaccarino, A.; Formari, G.; Saitta, M.; Prandoni, P. Presence of low titre of antiphospholipid antibodies in cancer patients: A prospective study. *Intern. Emerg. Med.* **2009**, *4*, 491–495. [CrossRef] [PubMed]
27. Vassalo, J.; Spector, N.; de Meis, E.; Rabello, L.S.; Rosolem, M.M.; do Brasil, P.E.; Salluh, J.I.; Soares, M. Antiphospholipid antibodies in critically ill patients with cancer: A prospective cohort study. *J. Crit. Care* **2014**, *29*, 533–538. [CrossRef] [PubMed]

Review

Current Controversies in Lupus Anticoagulant Detection

Gary W. Moore

Diagnostic Haemostasis & Thrombosis Laboratories, Department of Haemostasis and Thrombosis, Viapath Analytics, Guy's & St. Thomas' NHS Foundation Hospitals Trust, 4th floor North Wing, St. Thomas' Hospital, Westminster Bridge Road, London SE1 7EH, UK; gary.moore@viapath.co.uk; Tel.: +44-020-7188-0814; Fax: +44-020-7188-2726

Academic Editor: Ricard Cervera
Received: 27 September 2016; Accepted: 18 November 2016; Published: 2 December 2016

Abstract: Antiphospholipid syndrome is an autoimmune, acquired thrombophilia diagnosed when vascular thrombosis or pregnancy morbidity are accompanied by persistent antiphospholipid antibodies. Lupus anticoagulants (LA) are one of the criteria antibodies but calibration plasmas are unavailable and they are detected by inference based on antibody behaviour in a medley of coagulation-based assays. Elevated screening tests suggest the presence of a LA, which is confirmed with mixing tests to evidence inhibition and confirmatory tests to demonstrate phospholipid-dependence. At least two screening tests of different principle must be used to account for antibody heterogeneity and controversy exists on whether assays, in addition to dilute Russell's viper venom time and activated partial thromboplastin time, should be employed. A variety of approaches to raw data manipulation and interpretation attract debate, as does inclusion or exclusion of mixing studies in circumstances where the presence of a LA is already evident from other results. Therapeutic anticoagulation compromises coagulation-based assays but careful data interpretation and use of alternative reagents can detect or exclude LA in specific circumstances, and this aspect of LA detection continues to evolve. This review focuses on the main areas of debate in LA detection.

Keywords: activated partial thromboplastin time; antiphospholipid antibodies; antiphospholipid syndrome; dilute prothrombin time; dilute Russell's viper venom time; lupus anticoagulant; mixing tests; Taipan snake venom time

1. Introduction

Antiphospholipid syndrome (APS) is diagnosed when laboratory assays demonstrate the presence of persistent antiphospholipid antibodies (aPL) in patients presenting with thrombosis or pregnancy morbidity [1]. Crucially, thrombosis and pregnancy morbidity are by no means specific to APS and diagnosis is highly reliant on accurate and timely detection of aPL. Solid phase assays are employed to detect two of the criteria antibodies, anticardiolipin antibodies (aCL) and anti-β2-glycoprotein I antibodies (aβ2GPI), whilst lupus anticoagulants (LA) are detected in coagulation assays.

Standardisation difficulties for aPL assays persist, arising from issues such as antibody heterogeneity, reagent variability and differing interpretation strategies, and so generation of gold standard assays and reference plasmas remains elusive [2,3]. Whilst aCL and aβ2GPI assays can be calibrated to generate quantitative results, the presence of LA is inferred based on antibody behaviour in a medley of phospholipid-dependent coagulation assays [4–6].

No single type of coagulation test is sensitive to all LA and two test systems of differing analytical principles should be employed to maximise detection rates [4–6]. Classically, the medley for each test type comprises:

(a) a screening test that employs a low phospholipid concentration to accentuate the effect of LA by increasing competition with activated coagulation factors for limited phospholipid-binding sites

(b) performance of the screening test on a 1:1 mixture of index and normal pooled plasma (NPP) to evidence inhibition

(c) confirmatory tests that recapitulate the screening test but with concentrated phospholipid to partially or wholly swamp/overwhelm any LA, thereby demonstrating phospholipid dependence

A patient with a LA and no other causes of elevated clotting times present would be expected to generate elevated clotting times in the screening test and mixing test, and a significantly shorter clotting time with the confirmatory test that typically, but not always, returns into the reference range. As long as the composite for one of the test systems is consistent with the presence of a LA, you have found what you are looking for even if the other has given a normal screening test result. One of the problems with employing global coagulation assays to infer the presence of LA is that standard interpretation criteria necessarily assume that all else about the patient's coagulation status is normal, so each test has significant potential for compromised specificity, particularly in situations of therapeutic anticoagulation. This adds a further layer of complexity to LA identification and several guidelines with broad but not complete agreement are available to guide best practices [4–7]. The main interferences for each LA assay type are shown in Table 1.

Table 1. Main interfering factors in lupus anticoagulant assays affecting specificity.

Assay Types	Assays	Interferences
Intrinsic pathway-based assays	LA-responsive routine APTT Dilute APTT KCT SCT	*Non-LA causes of screening test elevation* Deficiencies of factors II, V, VIII, IX, X, XI, XII, PK, HMWK Reduced fibrinogen Anticoagulation with VKA, UFH, (LMWH), DFXa, DTI Non-phospholipid-dependent inhibitors *Shortening of screening test* Elevated FVIII, FIX Elevated fibrinogen
Extrinsic pathway-based assays	dPT ASLA	*Non-LA causes of screening test elevation* Deficiencies of factors II, V, X; (dPT only: factors VII, VIII, IX) Reduced fibrinogen Anticoagulation with VKA, DFXa, DTI, UFH Non-phospholipid-dependent inhibitors
Common pathway-based assays FX activation	dRVVT VLVT	*Non-LA causes of screening test elevation* Deficiencies of factors II, V, X Reduced fibrinogen Anticoagulation with VKA, DFXa, DTI, (UFH, LMWH) Non-phospholipid-dependent inhibitors
Common pathway-based assays FII activation	TSVT Textarin time	*Non-LA causes of screening test elevation* Deficiency of factor II; (Textarin time only: factor V) Reduced fibrinogen Anticoagulation with UFH, LMWH, DTI Non-phospholipid-dependent inhibitors

APTT, activated partial thromboplastin time; ASLA, activated seven lupus anticoagulant assay; DFXa, direct activated factor X inhibitor; dPT, dilute prothrombin time; dRVVT, dilute Russell's viper venom time; DTI, direct thrombin inhibitor; HMWK, high molecular weight kininogen; KCT, kaolin clotting time; LMWH, low molecular weight heparin; PK, prekallikrein; SCT, silica clotting time; TSVT, Taipan snake venom time; UFH, unfractionated heparin; VKA, vitamin K antagonist; VLVT, *Vipera lebetina* venom time.

2. Which Assays Should Be Used?

Numerous assay types for LA detection have been proposed and used over the years and earlier guidelines more or less gave practitioners free reign over which, and how many, to use, albeit with acknowledgement that the pairing of dilute Russell's viper venom time (dRVVT) and activated partial thromboplastin time (APTT) can achieve good detection rates [1,8–10]. However, the most recent guideline from the International Society on Thrombosis and Haemostasis (ISTH), published in late 2009, suggests that that the risk of false-positive results is increased to unacceptable levels if more than two screening tests are performed and restricts assay choice to only dRVVT, for its specificity to clinically significant antibodies [11], and APTT with low phospholipid concentration because of its sensitivity [4]. This recommendation has its basis in the considerable body of evidence indicating that the dRVVT and APTT pairing is diagnostically efficacious, and additionally, it serves to nurture common diagnostic practices. An important caveat here is that not all reagents from different manufacturers for the same test perform identically, particularly in the case of APTT [7,11–14], and reagents for LA detection must be chosen carefully. Many APTT reagents intended for routine coagulation screening to primarily detect coagulation factor deficiencies and monitor heparin therapy do not contain dilute phospholipid and/or a suitable phospholipid composition, compromising their responsiveness to LAs. Available routine APTT reagents span a continuum of low, intermediate and (relatively) high LA sensitivity [14], whilst other APTTs are specifically formulated only for use in LA detection [15,16]. A valuable recommendation is made in the guideline from the Clinical and Laboratory Standards Institute (CLSI) from 2014 to employ a LA-insensitive APTT in routine coagulation screening and a separate, LA-sensitive reagent when specifically investigating for LA [6,17]. This permits interpretation of LA assays themselves unencumbered by the possibility of many interfering factors if the routine APTT is normal [6,7]. However, considerations of resource availability, cost and convenience lead many diagnostic facilities to employ one APTT reagent for both routine and LA testing, which can adversely affect diagnostic outcomes [14]. It has been suggested that ellagic acid-activated APTTs are less sensitive to LA and that only silica-activated reagents should be used [4]. This is based on reports comparing routine APTT reagents employing different activators [13] yet ellagic acid-activated APTTs that are LA-sensitive have been described [18] and the low sensitivity in the other reagents is merely coincidental to their phospholipid composition [18,19]. Utility of dRVVT in LA detection is indisputable, to such an extent that some laboratories erroneously employ it in isolation for LA detection [20–22]. Aside from recognition that not all LA will manifest in dRVVT, hence recommendations in all guidelines to employ at least two different tests [4–7,19], between-reagent variation exists for dRVVT reagents too [23–26], albeit to a lesser extent than APTT.

The issue of whether extending LA detection repertoires to include additional tests inevitably increases false-positivity rates is contentious [5–7,19]. Certainly, there will always be natural statistical outliers for any test regardless of the statistical model applied for cut-off generation, so in that sense, the more screening tests you employ the risk of false-positivity does indeed increase. However, any elevated screening test will subsequently receive mix and confirm tests, so weak LA that might be missed due to the adoption of a higher cut-off will not go undetected, and any genuine outliers will generate concordant confirm results and negative mix results. It is the composite that secures diagnosis, not an isolated screening test result, and it is for this reasoning that guidelines from the CLSI [6] and the latest from British Committee for Standards in Haematology (BCSH) [5] do not exclude performance of tests additional to dRVVT and APTT.

An important consideration regarding other tests is that assays such as dilute prothrombin time (dPT), activated seven lupus anticoagulant assay (ASLA), textarin time and Taipan snake venom time (TSVT) have been shown to detect small numbers of LA that do not manifest in dRVVT or APTT, or at least the versions of those two assays employed in a given study, as well as antibodies that do manifest in dRVVT or APTT [7,19,27–32]. Epitope specificity variation, even for antibodies to domain I of β2glycoprotein I [3,33], means that the dRVVT and APTT combination alone cannot deliver diagnostic certainty. Alternative assays are commonly evaluated against one or both of those assays, creating a selection bias, and are thus disadvantaged in the context of perceived utility because they may be sensitive to different antibody subpopulations. Antibodies unreactive in dRVVT and APTT can be clinically significant [27–30,34–37], yet it is probably impractical and unnecessary for all laboratories to adopt extended LA-assay repertoires. Nonetheless, there is a case for use of additional assays in select patients or circumstances [5–7]. It could be reasonably argued that this additional testing should be the remit of reference laboratories, particularly for less familiar assays, although recent reports have suggested that addition of the relatively straightforward dPT to a dRVVT and APTT pairing can improve detection rates [7,19,28,37]. Concern has been expressed about lack of standardisation with some of the alternative assays [4], which is not unwarranted for assays such as kaolin clotting time [6,7,19], but not insurmountable for assays such as dPT and TSVT where available reagents coupled with suitable expertise and experience can render them clinically valuable [28,29,32,34,37].

3. Raw Data Manipulation

The end-point for all LA assays is of course a clotting time in seconds. We are all familiar with reporting and interpreting routine prothrombin times and APTTs in seconds and some practitioners prefer to do so with LA assays [22,25,38], although all guidelines recommend conversion of screen, mix and confirm results to normalised ratios [4–7]. This practice can reduce intra-assay, inter-assay and between-laboratory variation by minimising operator and/or analyser variability, accounting for reagent quality and stability issues, and NPP clotting time variation [6,7,39,40]. The controversy concerns whether the NPP clotting time or reference interval (RI) mean clotting time should comprise the denominator. The advantage of the former is that NPP is analysed alongside patient samples, thereby accounting for innate between-run analytical variation by reflecting operator/reagent/analyser performance in real time. The disadvantage is that not all NPP generate the same clotting times with different reagents for the same test type, which can systematically bias calculated ratios towards false-positive or false-negative results if the NPP value is towards an extreme of the RI and distant from the RI mean [6,7,19,40,41]. Adopting the RI mean circumvents this potentially serious problem, but requires generation of a robust RI, and NPP samples should continue to be run alongside test samples to reflect real time assay performance and identify sudden analytical difficulties [7,14].

Another gain from normalisation is that clotting times for confirmatory tests are often shorter than those for screening tests, even in NPP [13,41], which risks interpreting a screen and confirm discordance as a LA when it is merely a function of reagent properties. Normalising the clotting times of donor plasmas in RI generation virtually abolishes these discrepancies such that screen and confirm RIs are almost identical [13,41] and thus permitting direct comparison of screen and confirm results in patient samples.

4. Generation of Reference Intervals/Cut-Offs

The plethora of available reagents and analysers means that common RIs and cut-offs cannot be applied to any given test and that locally derived reference ranges are necessary [4–7,23,39–42]. Historically, RIs for LA assays have been derived parametrically from the RI mean ±2 standard deviations (SD) since normal donor population data for clotting tests are commonly Gaussian, or can be made so by data transformation [5–7,13,43]. The RI upper limit operates as the cut-off for determining screening test positivity and initiating mixing and confirmatory tests, whilst the RI mean clotting time can be employed to generate normalised ratios.

Whilst common practice for these and many other assays, the consequent upper limit cut-off at the 97.5th percentile results in 2.5% of observations being above that cut-off and representing false-positive screening tests. To reduce this possibility, the current ISTH guideline recommends application of the 99th percentile as the cut-off, which equates to the RI mean +2.3 SD for normally distributed data. This can reduce frequency of false positive results and thereby increase specificity, yet it is a statistical inevitability that this will reduce sensitivity [5–7]. The recommendation has proven controversial, partly because composite testing will identify any false-positive screening results and the slight reduction in sensitivity is therefore avoidable, but also because it is indicated that a 99th percentile value can be derived from a minimum of 40 donors when a minimum of 120 has been previously recommended [44]. Non-normally distributed data require considerably more. Sourcing that many normal donors is problematic and impractical for most diagnostic facilities and advice in the BCSH and CLSI guidelines maintains that generating a RI from its mean ±2 SD remains a valid and achievable proposition, and can be done with as few as 20 donors through validation exercises of previous cut-offs. The theoretical loss of sensitivity when adopting 99th percentile cut-offs has been substantiated in recent studies [45–47] and some workers continue to adopt 97.5th percentile, and even 95th percentile [41,46–48].

5. Confirmatory Tests

Despite recommendations in earlier as well as current guidelines, many laboratories have limited their APTT testing to screen and mix tests only, further restricting specificity of that assay system [22,49]. Indeed, the design and interpretation of the previously widely used kaolin clotting time, an APTT modification, is predicated on mixing test results [1,10]. This was partly due to poor availability of paired APTT confirm reagents [6]. Reports that pairing a known LA-sensitive APTT with an known insensitive reagent where practitioners ostensibly just need to perform two routine APTTs [14,17,48] have led to a greater but not complete uptake of APTT confirm assays [20–22,49]. Another approach is to screen with a LA-sensitive reagent, and if elevated, proceed to testing with the Staclot® LA assay. This involves performing APTTs on equal volume mixtures of index plasma and NPP with a separate, highly LA-sensitive reagent in the absence or presence of hexagonal phase phosphatidylethanolamine, the latter being the confirmatory test [6,15].

Antibody heterogeneity makes it crucial that confirmatory tests are based on whichever screening test is elevated [4–10]. Finding an elevated APTT screening test but attempting to discern phospholipid dependence by performing dRVVT screen and confirm assays will result in failure to accurately identify a LA that is dRVVT-unreactive.

Confirmation of phospholipid dependence is achieved by mathematically establishing that any difference between screen and confirm values is significant and due to more than analytical variance. Although the BCSH guidelines from 1991, 2000 and 2012 recommend the percent correction of screen ratio by confirm ratio for dRVVT, where >10% correction is considered significant, a variety of other calculations to assess for a significant difference have been proposed and used [9,50]. The issue was not directly addressed by ISTH guidelines until the 2009 update, where percent correction is also the recommendation, accompanied by a recommendation to assess the cut-off locally [4,8,10]. However, regulatory issues in some countries restrict practice to directions given by manufacturers in their package inserts and a common instruction is to calculate a ratio from dividing the screen result by the confirm result and assess against a cut-off [6,25]. Clearly, a high ratio is indicative of phospholipid dependence, yet this is often derived from raw clotting times without prior screen and confirm normalisation, which risks misclassification due to reagent-induced differences, and reduction in inter-method and inter-laboratory agreement [6,13,25]. In recognition, all current guidelines indicate that laboratories adopting this approach should first normalise screen and confirm results [4–7]. Some APTT-based assays, such as Staclot® LA, are assessed with deltas [6,15].

An interesting consequence of the availability of paired screen and confirm reagents is the adoption of so-called integrated testing where the reagent pair are assayed in parallel with each patient [4,7,11,25,48]. This permits immediate assessment for phospholipid dependence and circumvents the traditional algorithm of performing a mixing test in response to an elevated screening test and initiation of a confirmatory test if the mix is also elevated. One potential benefit is improved detection of weaker LA where the prolongation of a patient's basal screening test clotting time is insufficient to exceed the cut-off yet the screen and confirm discordance elevates the ratio, or indicates high percentage correction, and reveals the antibody [51–53]. The immediate availability of demonstration of phospholipid dependence has led some to question the role of mixing tests in the LA detection armoury [6,7,49,51,53–57], to which we will now turn our attention.

6. Mixing Tests

Lupus anticoagulants are, by definition, in vitro inhibitors, so it is entirely logical to investigate an elevated screening test with a mixing test in the first instance. If an elevated screening test does not prove to be due to an inhibitor it wastes time and resources to perform the confirmatory test. Or does it? Whilst that was accepted wisdom and standard practice for some time it is now widely accepted, and acknowledged in guidelines, that mixing tests introduce a dilution factor that can generate false-negative results such that adopting the traditional screen-mix-confirm algorithm may reduce detection rates [4–7,51,57–60]. This has fuelled acceptance and the adoption of the integrated testing model as a faster and more cost-effective LA detection strategy, yet it does not represent the full picture and returns us to the issue of specificity.

As stated in the BCSH guideline [5], a patient with clear screen and confirm discordance in undiluted plasma and no other causes of elevated clotting times but with a normal mixing test can be reasonably considered to have a LA, the mixing test result being merely due to a limitation of test design. Evidence for absence of a co-existing abnormality largely comes from the coagulation screen employing a LA-insensitive routine APTT, if normal, but also from the confirmatory test whose partner screening test was elevated, in that a normal confirmatory test is additional evidence that interfering factors specific to that test system are not present [6,19,54,61]. However, if significant screen and confirm discordance is apparent but the confirm result is elevated, consideration must be given to the possibility of a co-existing abnormality such as a factor deficiency or undisclosed therapeutic anticoagulation that could compromise LA detection [6,7,26,54,61,62]. Alternatively, some LA possess a degree of resistance to the swamping effect of confirm reagents and the confirm result is elevated for this reason alone [4,7,54]. In such circumstances, mixing tests can be invaluable by confirming the presence of inhibition or correcting for some non-LA abnormalities. Specificity can be further increased if a confirm mixing test is also performed as it aids discrimination between potent LA, LA + co-existing abnormality or non-LA abnormality [1,7,54,63,64]. Non-LA causes of elevated clotting times usually generate concordant screen and confirm ratios and integrated testing generates a normal ratio or low percent correction, thereby excluding a LA. However, exceptionally potent LA may also generate such results when tested with undiluted plasma and the dilution factor in mixing tests can be used to diagnostic advantage by dampening the effect of the antibody and reveal screen and confirm discordance [19,54,55,64]. The important message here is that some LA can be reliably detected without mixing tests and the decision point for progressing to mixing tests is whether or not the confirm result from undiluted plasma is elevated [6,7,19,54,61]. This has led the CLSI to recommend a re-prioritisation of testing order to screen-confirm-mix, the mixing tests only being performed when other tests are not clear cut and the decision is made on a case-by-case basis.

Assays with higher specificity tend to require less mixing tests to clearly demonstrate LA [56,65]. Although considered rare, it is inevitable that the LA cofactor effect cannot be detected when mixing tests are omitted [7,55]. The LA cofactor effect is the paradoxical further prolongation of the patient's clotting time upon mixing with NPP [6,10,55]. The phenomenon is thought to be due a patient's plasma being deficient in an as yet undefined cofactor that is essential for LA to exert their in vitro anticoagulant effect. The NPP normalises the cofactor level, thereby permitting greater expression of the LA-induced inhibition. The cofactor has been proposed but not proven to be prothrombin or β2-glycoprotein I [6].

Accepting that mixing tests maintain a valuable role in LA detection but are compromised by the dilution effect necessitates exploration of strategies to maximise detection rates. Adopting dilutions that favour an excess of index plasma has been proposed but does not guarantee correction of severe or multiple factor deficiencies [6,10], and although potentially useful in detecting weaker antibodies [66], it is resource expensive and practically cumbersome to perform multiple mixing test dilutions. Instead, efforts have been centred on alternative approaches to interpretation of 1:1 mixing tests. It is common to interpret mixing tests for standard coagulation tests against the RI for undiluted plasma [67] but mixing test-specific RIs for LA assays have been shown to be narrower than for undiluted plasma due to clotting times of normal samples at the extremes being compensated upon mixing with NPP [6,7,11,13,60]. The narrower RI and thus lower cut-off increases sensitivity for inhibition [6,13,38,60,68] and is now recommended for interpreting LA mixing tests [4,6,7]. The index of circulating anticoagulant (ICA) calculation [46,69] is an alternative recommendation [4,6], although recent studies have reported that a mixing test-specific cut-off is more sensitive than ICA in detection of inhibition [38,68].

7. Testing Anticoagulated Patients

All therapeutic anticoagulants have potential to compromise LA testing with some or most available assays and laboratory investigation for LA is best postponed until after discontinuation of anticoagulant treatment [4–6]. Despite this, it is common for samples from anticoagulated patients to be submitted for diagnostic LA testing [7,70]. It is therefore incumbent on diagnostic laboratories to recognise and maximise situations where LA can nonetheless be detected, yet be honest with themselves and their service users when they cannot. This would normally take the form of performing the assays on each patient and interpreting in light of assay design and limitations, degree of anticoagulation, reagent properties and whether strategies such as mixing studies employing both screen and confirm assays, and heparin neutralisers, have accounted for anticoagulation interference.

7.1. Vitamin K Antagonists

An almost perennial controversy is whether LA assays performed on undiluted plasma from patients receiving vitamin K antagonist (VKA) anticoagulation are reliable, particularly dRVVT [57,71–74]. Some contend that the multiple acquired factor deficiency of VKA therapy does indeed compromise LA detection [71,73], whilst others maintain that any screen and confirm discordance is reliable despite any confirm test elevation [51,72]. Isert et al. proposed an elevation of the dRVVT normalised screen/confirm ratio cut-off from 1.3 to 1.7 when testing plasma from VKA anticoagulated patients (International Normalised Ratio (INR) 2.0–3.0), which improved accuracy in known APS patients but risked a mild reduction in sensitivity [74]. In view of the controversy, others choose to disregard results from undiluted plasma and instead rely on mixing tests where an elevated screen evidences inhibition, and a reduced and often normal confirm evidences that the mix corrects the VKA effect and the reagent corrects the LA [1,5,6,63]. A positive result is diagnostic but negative mixing tests are inconclusive due to the dilution effect [1,5,6].

An additional tool for detecting LA in VKA anticoagulated patients is use of assays based on 'VKA-insensitive' snake venom prothrombin activators [31,34]. Textarin and Taipan venoms are phospholipid-dependent, so diluting the phospholipid makes the assays LA-responsive, and employing prothrombin activators aids specificity. In place of concentrated phospholipid confirmatory tests, both venoms are commonly paired with ecarin venom, which contains a phospholipid-independent, 'VKA-insensitive' prothrombin activator [31,32,34,63,75]. Combining TSVT and ecarin time (ET) with dRVVT and APTT mixing tests increases detection rates, partly because antibodies 'lost' by dilution can still manifest in TSVT/ET [5,6,63], and also because TSVT will detect a small proportion of LA unreactive with dRVVT and APTT [32,34,63,76].

It has been suggested that that there are no standardised commercial assays employing these venoms [4], and whilst true for Textarin, paired TSVT and ET reagents have been available for some time from at least one manufacturer [7,32,34]. The ISTH scientific sub-committee chose not to recommend these assays for use in VKA anticoagulated patients because it was considered they required further critical evaluation in that setting [4]. Both BCSH and CLSI guidelines do however suggest they can be used in this patient population, possibly because authors on those committees had more direct personal experience of them [5–7,34,63,76].

An important consideration for patients with LA who are treated with VKA anticoagulation is that they are monitored with a phospholipid dependent test, the prothrombin time (PT), to generate the INR. Fortunately, the high phospholipid concentration in thromboplastin reagents means that >95% of patients with APS have a normal PT in the absence of other coagulopathies [5]. For patients whose LA does prolong the locally employed PT prior to anticoagulation, thereby risking overestimation of anticoagulation, monitoring with an alternative, LA-insensitive thromboplastin shown to generate a normal baseline for that patient is usually achievable. More rarely, amidolytic factor X assays can be employed [1,5]. Reagents comprising recombinant tissue factor and purified phospholipids tend to be slightly more prone to LA interference. Additionally, point-of-care devices for INR generation can also be affected by LA and baseline PT on the device should be performed prior to commencement of anticoagulation [5].

7.2. Heparins

Interference by unfractionated heparin (UFH) is inevitable in LA assays and attempting to detect the antibodies in this situation is largely discouraged [4–7]. However, most commercial dRVVT reagents contain heparin neutralisers that are effective up to a specified UFH level, commonly between 0.8 and 1.0 U/mL. The important question for practitioners who interpret these results is how they know the neutraliser has quenched the UFH without simultaneously assaying UFH levels. Once again, the confirmatory test comes to our rescue. An elevated screen can be attributed to a LA if the confirm ratio is normal since its phospholipid has corrected the LA and the neutraliser has dealt with the UFH [6]. Assays employing confirmatory tests derived from platelet material should not be used as the platelet factor 4 will neutralise UFH and can generate false-positive screen and confirm discordance [5,6,77].

Low molecular weight heparins (LMWH) are generally considered to have little or no effect on standard prothrombin time and APTT assays yet dose, LMWH type and reagent variability can generate elevated clotting times [78,79]. Although interference of LMWH in LA assays is less frequent than with some other anticoagulants it does occur and the presence of LMWH must be taken into account when interpreting LA assays, even in reagents with heparin neutralisers [6,62,80].

7.3. Direct Oral Anticoagulants

The direct oral anticoagulants (DOAC) inevitably complicate LA testing and interpretation, and data continue to emerge concerning their effects on different LA assays [62,80–84]. Numerous studies have reported greater elevation of dRVVT screen values than confirm values with direct factor Xa (DFXa) inhibitors in non-LA patients or spiked NPP [26,62,81,83–87]. The screen and confirm discordance is sufficient to generate almost ubiquitous false-positivity at peak rivaroxaban concentrations, and to a lesser extent at trough levels [82]. Similar patterns are seen with edoxaban but to a lesser extent with apixaban [81,86]. Being direct inhibitors, mixing tests are also elevated, further increasing the risk of false-positive interpretations. As might be expected, APTT testing for LA is less often affected by DFXa inhibitors but caution is required when interpreting results as reagent variability exists [62,81,86,87].

These complications can be overcome with the TSVT/ET pairing since the direct prothrombin activation by both venoms bypasses the effects of DFXa inhibitors. Recent studies have reported successful LA detection with TSVT/ET in patients on rivaroxaban [75,84,88,89]. However, TSVT/ET alone will not detect all LA and additionally employing APTT-based assays that are insensitive to DFXa inhibitors [81] should help account for antibody heterogeneity. Newly developed dRVVT reagents that are less affected by warfarin and rivaroxaban, and thus exhibiting improved specificity, have recently become available and been evaluated [26,81,88]. Although both therapeutic anticoagulants still elevate the screen values, patients without LA generate concordant confirm values and the false-positive interpretations encountered with most other dRVVT reagents do not ensue. Another recent study reported similar findings with home-brew dRVVT reagents [84]. There does appear to be a slight loss of sensitivity with the commercial reagents, albeit no less sensitive than some other commercially available dRVVT reagents [25,26,88]. In view of this, a panel of DOAC-insensitive APTT, VKA and DOAC insensitive dRVVT and TSVT/ET could maximise detection rates in anticoagulated patients.

Being a direct thrombin inhibitor, dabigatran potentially interferes with all LA tests and false positive interpretations are common, including elevations of mixing tests [62,80,82,90–92]. Testing for LA is best postponed until dabigatran is withdrawn, although a recent study reported successful in vitro idarucizumab-induced reversal of dabigatran anticoagulation in dRVVT and other routine coagulation tests [92].

8. Can Lupus Anticoagulants Be Quantified?

The current detection of LA by inference in coagulation assays provides evidence of presence but not concentration. The ability to quantify LA could help stratify patients into risk groups but is complicated by the lack of a true plasma standard with assigned activity since the antibodies are heterogeneous [93]. This heterogeneity, coupled with between-reagent variability, can result in a given LA appearing to be a strong positive with one reagent but weak or even negative with another [12,17,20,22,50], making even semi-quantitative assessment of results as weak, moderate or strong almost meaningless.

An early attempt to quantify LA involved performing APTT and dRVVT screen and confirm assays on 1:1 mixtures of test plasma and NPP. The ratio between the two clotting times in each test was divided by the corresponding ratio for the NPP itself, generating a third ratio referred to as the Lupus Ratio [94]. The upper limit of a normal reference population was deemed to represent one LA Unit (LA-U) and dilutions of a single 'strong' LA-positive plasma used to construct calibration curves. Although exhibiting high sensitivity and specificity, use of a single plasma for calibration was a limitation and the Lupus Ratio alone was later proposed as a semi-quantitative procedure [95].

Perhaps unsurprisingly, other workers have explored spiking NPP with antibodies known to possess LA activity. Le Querrec et al. demonstrated that various clotting assays could be calibrated against a NPP spiked with monoclonal antibodies to the two most common antigenic targets of LA, β2-glycoprotein I and prothrombin [96]. However, there were large differences in responsiveness between the assays and reagents employed, and the two monoclonal antibodies would not necessarily account for the polyclonal nature of LA or other antigenic targets such as annexin A5 [97].

Tripodi et al. performed a feasibility study for an alternative approach involving assignment of a LA sensitivity index (LASI) to LA assay reagents directly analogous to the international sensitivity index used for INR determination [98]. The usual ratios generated from LA testing were converted into a new 'universal' scale called standardised LA-ratio (SLA-ratio). Reagent differences manifested when SLA-ratio was calculated from LASI calibration with NPP spiked with purified IgG from patients with 'strong' LA and aβ2GPI but were abrogated when calibration was against a set of plasmas from LA-positive patients.

Possibly the most promising approach has been to generate a ratio from the peak height and lag time thrombin generation parameters and calibrate against NPP spiked with monoclonal antibodies to β2-glycoprotein I and prothrombin to quantify LA activity in arbitrary units [93]. Although the above mentioned limitations with this calibration material remain, combining the quantified LA results with those of other assays, such as the solid-phase assay aβ2GPI titre and factor VII activity, permitted a layered strategy for thrombotic risk assessment, but which was not possible with the thrombin generation parameters alone.

Can lupus anticoagulants be quantified? Not quite, but we are getting ever closer, although generating that polyclonal plasma standard applicable to all patients and antigens remains a high hurdle.

9. Conclusions

The publication of recent guideline updates by the ISTH and BCSH, and the new CLSI guideline, have gone a long way towards harmonising diagnostic practices, despite incomplete agreement on certain issues [7,19,37,40,45,65,70]. Most laboratories will likely continue to employ only the dRVVT and APTT pairing, which will detect most LA, yet experienced proponents of other assays are unlikely to be deterred from using them as they have evidenced genuine clinical utility, at least in certain circumstances [28,34,37,75,84]. Although the mixing test debate will continue apace [7,13,49,51,54–57,59,61], many diagnostic departments have already made their decision whether to retain this member of the medley or not [45,49,56,57,61,63,68], despite advice that it is invaluable in some situations and probably unnecessary in others. Until the unlikely event that LA calibration plasmas become available, practitioners will continue to apply their preferred statistical model for cut-off generation with respect to the balance between sensitivity and specificity. Consensus has almost been reached on how to evidence phospholipid dependence, with most laboratories applying percent correction or normalised screen/confirm ratio. Similarly for mixing tests, although recent reports suggest mixing test cut-off is more sensitive than ICA in detecting inhibition in multiple assays [38,68,99] and additional studies may consolidate this view sufficient for a firm recommendation in future guidelines.

Detection of LA in anticoagulated patients is possible in many cases upon careful application of appropriate assays and interpretation strategies. It is conceivable that DOAC and VKA insensitive reagents, and/or addition of DOAC reversal agents, will greatly improve this situation in the near future.

Recognition of the importance of antibodies to domain I of β_2-glycoprotein I in the pathophysiology of APS [33,100], and recent availability of solid phase assays to detect them [101], may have posed a threat to the continued use of coagulation assay medleys to detect clinically significant aPL [7]. However, antibody heterogeneity, proven clinical utility, and detection of clinically significant LA without β_2-glycoprotein I specificity will keep coagulation-based LA assays in diagnostic repertoires for the foreseeable future.

Conflicts of Interest: The author declares no conflict of interest.

References

1. Greaves, M.; Cohen, H.; Machin, S.J.; Mackie, I. Guidelines on the investigation and management of the antiphospholipid syndrome. *Br. J. Haematol.* **2000**, *109*, 704–715. [CrossRef] [PubMed]
2. Devreese, K.M. Standardization of antiphospholipid antibody assays. Where do we stand? *Lupus* **2012**, *21*, 718–721. [CrossRef] [PubMed]
3. De Groot, P.G.; Urbanus, R.T. The future of antiphospholipid antibody testing. *Semin. Thromb. Hemost.* **2012**, *38*, 412–420. [CrossRef] [PubMed]
4. Pengo, V.; Tripodi, A.; Reber, G.; Rand, J.H.; Ortel, T.L.; Galli, M.; De Groot, P.G. Subcommittee on Lupus Anticoagulant/Antiphospholipid Antibody of the Scientific and Standardisation Committee of the International Society on Thrombosis and Haemostasis. Update of the guidelines for lupus anticoagulant detection. *J. Thromb. Haemost.* **2009**, *7*, 1737–1740. [CrossRef] [PubMed]
5. Keeling, D.; Mackie, I.; Moore, G.W.; Greer, I.A.; Greaves, M. British Committee for Standards in Haematology. Guidelines on the investigation and management of antiphospholipid syndrome. *Br. J. Haematol.* **2012**, *157*, 47–58. [CrossRef] [PubMed]
6. Clinical and Laboratory Standards Institute (CLSI). *Laboratory Testing for the Lupus Anticoagulant: Approved Guideline*; CLSI Document H60-A; Clinical and Laboratory Standards Institute: Wayne, PA, USA, 2014.
7. Moore, G.W. Recent guidelines and recommendations for laboratory detection of lupus anticoagulants. *Semin. Thromb. Hemost.* **2014**, *40*, 163–171. [CrossRef] [PubMed]
8. Exner, T.; Triplett, D.A.; Taberner, D.; Machin, S.J. Guidelines for testing and revised criteria for lupus anticoagulants. SSC Subcommittee for the Standardization of Lupus Anticoagulants. *Thromb. Haemost.* **1991**, *65*, 320–322. [PubMed]
9. Machin, S.J.; Giddings, J.C.; Greaves, M.; Hutton, R.A.; Mackie, I.J.; Malia, R.G.; Taberner, D.A. Guidelines on testing for the lupus anticoagulant. Lupus Anticoagulant Working Party on behalf of the BCSH Haemostasis and Thrombosis Task Force. *J. Clin. Pathol.* **1991**, *44*, 885–889.
10. Brandt, J.T.; Triplett, D.A.; Alving, B.; Scharrer, I. Criteria for the diagnosis of lupus anticoagulants: An update. On behalf of the Subcommittee on Lupus Anticoagulant/Antiphospholipid Antibody of the Scientific and Standardisation Committee of the ISTH. *Thromb. Haemost.* **1995**, *74*, 1185–1190. [PubMed]
11. Tripodi, A. Laboratory testing for lupus anticoagulants: A review of issues affecting results. *Clin. Chem.* **2007**, *53*, 1629–1635. [CrossRef] [PubMed]
12. Denis-Magdelaine, A.; Flahault, A.; Verdy, E. Sensitivity of sixteen APTT reagents for the presence of lupus anticoagulants. *Haemostasis* **1995**, *25*, 98–105. [CrossRef] [PubMed]

13. Kershaw, G.; Suresh, S.; Orellana, D.; Nguy, Y.M. Laboratory identification of lupus anticoagulants. *Semin. Thromb. Hemost.* **2012**, *38*, 375–384. [CrossRef] [PubMed]
14. Fritsma, G.A.; Dembitzer, F.R.; Randhawa, A.; Marques, M.B.; Van Cott, E.M.; Adcock-Funk, D.; Peerschke, E.I. Recommendations for appropriate activated partial thromboplastin time reagent selection and utilization. *Am. J. Clin. Pathol.* **2012**, *137*, 904–908. [CrossRef] [PubMed]
15. Charles, L.A.; McGlasson, D.L.; Hawksworth, B.A.; Ashcraft, J.H.; Ortel, T.L. Evaluation of a modified procedure for Staclot LA for the confirmation of lupus anticoagulants. *Blood Coagul. Fibrinolysis* **1994**, *5*, 601–604. [PubMed]
16. Devreese, K.M. Evaluation of a new silica clotting time in the diagnosis of lupus anticoagulants. *Thromb. Res.* **2007**, *120*, 427–438. [CrossRef] [PubMed]
17. Dembitzer, F.R.; Suarez, Y.; Aledort, L.M.; Peerschke, E.I. Screening coagulation testing using the APTT: Which reagent to choose? *Am. J. Hematol.* **2010**, *85*, 726–727. [CrossRef] [PubMed]
18. Kumano, O.; Ieko, M.; Naito, S.; Yoshida, M.; Takahashi, N. APTT reagent with ellagic acid as activator shows adequate lupus anticoagulant sensitivity in comparison to silica-based reagent. *J. Thromb. Haemost.* **2012**, *10*, 2338–2343. [CrossRef] [PubMed]
19. Moore, G.W. Commonalities and contrasts in recent guidelines for lupus anticoagulant detection. *Int. J. Lab. Haematol.* **2014**, *36*, 364–373. [CrossRef] [PubMed]
20. Jennings, I.; Greaves, M.; Mackie, I.J.; Kitchen, S.; Woods, T.A.; Preston, F.E. UK National External Quality Assessment Scheme for Blood Coagulation. Lupus anticoagulant testing: Improvements in performance in a UK NEQAS proficiency testing exercise after dissemination of national guidelines on laboratory methods. *Br. J. Haematol.* **2002**, *119*, 364–369. [CrossRef] [PubMed]
21. Moffat, K.A.; Ledford-Kraemer, M.R.; Plumhoff, E.A.; McKay, H.; Nichols, W.L.; Meijer, P.; Hayward, C.P. Are laboratories following published recommendations for lupus anticoagulant testing? An international evaluation of practices. *Thromb. Haemost.* **2009**, *101*, 178–184. [CrossRef] [PubMed]
22. Dembitzer, F.R.; Ledford Kraemer, M.R.; Meijer, P.; Peerschke, E.I. Lupus anticoagulant testing: Performance and practices by North American clinical laboratories. *Am. J. Clin. Pathol.* **2010**, *134*, 764–773. [CrossRef] [PubMed]
23. Lawrie, A.S.; Mackie, I.J.; Purdy, G.; Machin, S.J. The sensitivity and specificity of commercial reagents for the detection of lupus anticoagulant show marked differences in performance between photo-optical and mechanical coagulometers. *Thromb. Haemost.* **1999**, *81*, 758–762. [PubMed]
24. Moore, G.W.; Savidge, G.F. Heterogeneity of Russell's viper venom affects the sensitivity of the dilute Russell's viper venom time to lupus anticoagulants. *Blood Coagul. Fibrinolysis* **2004**, *15*, 279–282. [CrossRef] [PubMed]
25. McGlasson, D.L.; Fritsma, G.A. Comparison of six dilute russell viper venom time lupus anticoagulant screen/confirm assay kits. *Semin. Thromb. Hemost.* **2013**, *39*, 315–319. [PubMed]
26. Depreter, B.; Devreese, K.M. Dilute Russell's viper venom time reagents in lupus anticoagulant testing: A well-considered choice. *Clin. Chem. Lab. Med.* **2016**. [CrossRef] [PubMed]
27. Liestøl, S.; Jacobsen, E.M.; Wisløff, F. Dilute prothrombin-time based lupus ratio test. Integrated LA testing with recombinant tissue thromboplastin. *Thromb. Res.* **2002**, *105*, 177–182. [CrossRef]
28. Mackie, I.J.; Lawrie, A.S.; Greenfield, R.S.; Guinto, E.R.; Machin, S.J. A new lupus anticoagulant test based on dilute prothrombin time. *Thromb. Res.* **2004**, *114*, 673–674.
29. Devreese, K.M.J. Evaluation of a new commercial dilute prothrombin time in the diagnosis of lupus anticoagulants. *Thromb. Res.* **2008**, *123*, 404–411. [CrossRef] [PubMed]
30. Moore, G.W.; Smith, M.P.; Patel, Y.; Savidge, G.F. The Activated Seven Lupus Anticoagulant (ASLA) assay: A new test for lupus anticoagulants (LAs). Evidence that some LAs are detectable only in extrinsic pathway based assays. *Blood Coagul. Fibrinolysis* **2002**, *13*, 261–269. [CrossRef] [PubMed]
31. Triplett, D.A.; Stocker, K.F.; Unger, G.A.; Barna, L.K. The Textarin/Ecarin ratio: A confirmatory test for lupus anticoagulants. *Thromb. Haemost.* **1993**, *70*, 925–931. [PubMed]
32. Moore, G.W.; Culhane, A.P.; Maloney, J.C.; Archer, R.A.; Breen, K.A.; Hunt, B.J. Taipan snake venom time coupled with ecarin time enhances lupus anticoagulant detection in non-anticoagulated patients. *Blood Coagul. Fibrinolysis* **2016**, *27*, 477–480. [CrossRef] [PubMed]
33. De Groot, P.G.; Urbanus, R.T. The significance of autoantibodies against β2-glycoprotein I. *Blood* **2012**, *120*, 266–274. [CrossRef] [PubMed]

34. Moore, G.W.; Smith, M.P.; Savidge, G.F. The Ecarin time is an improved confirmatory test for the Taipan snake venom time in warfarinised patients with lupus anticoagulants. *Blood Coagul. Fibrinolysis* **2003**, *14*, 307–312. [CrossRef] [PubMed]

35. Galli, M.; Borrelli, G.; Jacobsen, E.M.; Marfisi, R.M.; Finazzi, G.; Marchioli, R.; Wisløff, F.; Marziali, S.; Morboeuf, O.; Barbui, T. Clinical significance of different antiphospholipid antibodies in the WAPS (warfarin in the antiphospholipid syndrome) study. *Blood* **2007**, *110*, 1178–1183. [CrossRef] [PubMed]

36. Moore, G.W.; Rangarajan, S.; Savidge, G.F. The activated seven lupus anticoagulant assay detects clinically significant antibodies. *Clin. Appl. Thromb. Haemost.* **2008**, *14*, 332–337. [CrossRef] [PubMed]

37. Swadzba, J.; Iwaniec, T.; Pulka, M.; De Laat, B.; De Groot, P.G.; Musial, J. Lupus anticoagulant: Performance of the tests as recommended by the latest ISTH guidelines. *J. Thromb. Haemost.* **2011**, *9*, 1776–1783. [CrossRef] [PubMed]

38. Depreter, B.; Devreese, K.M. Differences in lupus anticoagulant final conclusion through clotting time or Rosner index for mixing test interpretation. *Clin. Chem. Lab. Med.* **2016**, *54*, 1511–1516. [CrossRef] [PubMed]

39. Pradella, P.; Azzarini, G.; Santarossa, L.; Caberlotto, L.; Bardin, C.; Poz, A.; D'Aurizio, F.; Giacomello, R. Cooperation experience in a multicentre study to define the upper limits in a normal population for the diagnostic assessment of the functional lupus anticoagulant assays. *Clin. Chem. Lab. Med.* **2013**, *51*, 379–385. [CrossRef] [PubMed]

40. Krilis, S.A.; Giannakopoulos, B. Laboratory methods to detect antiphospholipid antibodies. *Hematol. Am. Soc. Hematol. Educ. Program* **2014**, *2014*, 321–328. [CrossRef] [PubMed]

41. Moore, G.W.; Brown, K.L.; Bromidge, E.S.; Drew, A.J.; Ledford-Kraemer, M.R. Lupus anticoagulant detection: Out of control? *Int. J. Lab. Haematol.* **2013**, *35*, 128–136. [CrossRef] [PubMed]

42. Gardiner, C.; Mackie, I.J.; Malia, R.G.; Jones, D.W.; Winter, M.; Leeming, D.; Taberner, D.A.; Machin, S.J.; Greaves, M. The importance of locally derived reference ranges and standardized calculation of dilute Russell's viper venom time results in screening for lupus anticoagulant. *Br. J. Haematol.* **2000**, *111*, 1230–1235. [CrossRef] [PubMed]

43. Gerbutavicius, R.; Fareed, J.; Messmore, H.L.; Iqbal, O.; Hoppensteadt, D.A.; Wehrmacher, W.H.; Demir, M.; Piccolo, P.; Ahmad, S.; Ma, Q.; et al. Reference intervals of the dilute tissue thromboplastin inhibition and dilute Russell's viper venom tests revisited. *Clin. Appl. Thromb. Hemost.* **2002**, *8*, 115–124. [CrossRef] [PubMed]

44. Clinical and Laboratory Standards Institute (CLSI). *Defining, Establishing, and Verifying Reference Intervals in the Clinical Laboratory: Approved Guideline*, 3rd ed.; CLSI Document EP28-A3c; Clinical and Laboratory Standards Institute: Wayne, PA, USA, 2008.

45. Martinuzzo, M.E.; Cerrato, G.S.; Varela, M.L.; Adamczuk, Y.P.; Forastiero, R.R. New guidelines for lupus anticoagulant: Sensitivity and specificity of cut-off values calculated with plasmas from healthy controls in mixing and confirmatory tests. *Int. J. Lab. Hematol.* **2012**, *34*, 208–213. [CrossRef] [PubMed]

46. Kumano, O.; Ieko, M.; Naito, S.; Yoshida, M.; Takahashi, N.; Aoki, T. Index of circulating anticoagulant cut-off value establishment in activated partial thromboplastin time mixing test for lupus anticoagulant diagnosis. *J. Thromb. Haemost.* **2013**, *11*, 1919–1922. [PubMed]

47. Averina, M.; Johannesen, S.; Brox, J. Diagnostic accuracy of silica clotting time method for lupus anticoagulant in a clinical population with various symptoms of antiphospholipid syndrome. *Lupus* **2016**, *25*, 418–422. [CrossRef] [PubMed]

48. Li, R.; Swaelens, C.; Vandermijnsbrugge, F.; Cantinieaux, B. Applying a direct aPTT ratio (PlatelinLS/ActinFS) permits to identify rapidly and reliably a bleeding-related factor deficiency or a lupus anticoagulant sequential to an isolated prolongation of aPTT in paediatric pre-operative screening. *Eur. J. Haematol.* **2016**, *96*, 578–585. [CrossRef] [PubMed]

49. Chandrashekar, V. Dilute Russell's viper venom and activated partial thromboplastin time in lupus anticoagulant diagnosis: Is mixing essential? *Blood Coagul. Fibrinolysis* **2016**, *27*, 408–411. [CrossRef] [PubMed]

50. Jennings, I.; Mackie, I.; Arnout, J.; Preston, F.E. UK National External Quality Assessment Scheme for Blood Coagulation. Lupus anticoagulant testing using plasma spiked with monoclonal antibodies: Performance in the UK NEQAS proficiency testing programme. *J. Thromb. Haemost.* **2004**, *2*, 2178–2184. [CrossRef] [PubMed]

51. Reber, G.; Meijer, P. In ECAT veritas? *Lupus* **2012**, *21*, 722–724. [CrossRef] [PubMed]

52. Jacobsen, E.M.; Wisløff, F. False negative screening tests for lupus anticoagulants—An unrecognised problem? *Thromb. Res.* **1996**, *82*, 445–451. [CrossRef]

53. Moore, G.W.; Henley, A.; Greenwood, C.K.; Rangarajan, S. Further evidence of false negative screening for lupus anticoagulants. *Thromb. Res.* **2008**, *121*, 477–484. [CrossRef] [PubMed]
54. Devreese, K.M.J. No more mixing tests required for integrated assay systems in the laboratory diagnosis of lupus anticoagulants. *J. Thromb. Haemost.* **2010**, *8*, 1120–1122. [CrossRef] [PubMed]
55. Tripodi, A. To mix or not to mix in lupus anticoagulant testing? That is the question. *Semin. Thromb. Hemost.* **2012**, *38*, 385–389. [CrossRef] [PubMed]
56. Hong, S.K.; Hwang, S.M.; Kim, J.E.; Kim, H.K. Clinical significance of the mixing test in laboratory diagnoses of lupus anticoagulant: The fate of the mixing test in integrated lupus anticoagulant test systems. *Blood Coagul. Fibrinolysis* **2012**, *23*, 739–744. [CrossRef] [PubMed]
57. Pennings, M.T.; De Groot, P.G.; Meijers, J.C.; Huisman, A.; Derksen, R.H.; Urbanus, R.T. Optimisation of lupus anticoagulant tests: Should test samples always be mixed with normal plasma? *Thromb. Haemost.* **2014**, *112*, 736–742. [CrossRef] [PubMed]
58. Male, C.; Lechner, K.; Speiser, W.; Pabinger, I. Transient lupus anticoagulants in children: Stepwise disappearance of diagnostic features. *Thromb. Haemost.* **2000**, *83*, 174–175. [PubMed]
59. Thom, J.; Ivey, L.; Eikelboom, J. Normal plasma mixing studies in the laboratory diagnosis of lupus anticoagulant. *J. Thromb. Haemost.* **2003**, *1*, 2689–2691. [CrossRef] [PubMed]
60. Moore, G.W.; Savidge, G.F. The dilution effect of equal volume mixing studies compromises confirmation of inhibition by lupus anticoagulants even when mixture specific reference ranges are applied. *Thromb. Res.* **2006**, *118*, 523–528. [CrossRef] [PubMed]
61. Devreese, K.M.; de Laat, B. Mixing studies in lupus anticoagulant testing are required at least in some type of samples. *J. Thromb. Haemost.* **2015**, *13*, 1475–1478. [CrossRef] [PubMed]
62. Martinuzzo, M.E.; Barrera, L.H.; D'adamo, M.A.; Otaso, J.C.; Gimenez, M.I.; Oyhamburu, J. Frequent false-positive results of lupus anticoagulant tests in plasmas of patients receiving the new oral anticoagulants and enoxaparin. *Int. J. Lab. Hematol.* **2014**, *36*, 144–150. [CrossRef] [PubMed]
63. Moore, G.W. Combining Taipan snake venom time/Ecarin time screening with the mixing studies of conventional assays increases detection rates of lupus anticoagulants in orally anticoagulated patients. *Thromb. J.* **2007**, *5*, 12. [CrossRef] [PubMed]
64. Favaloro, E.J.; Bonar, R.; Zebeljan, D.; Kershaw, G.; Marsden, K. Laboratory investigation of lupus anticoagulants: Mixing studies are sometimes required. *J. Thromb. Haemost.* **2010**, *8*, 2828–2831. [CrossRef] [PubMed]
65. Chantarangkul, V.; Biguzzi, E.; Asti, D.; Palmucci, C.; Tripodi, A. Laboratory diagnostic outcome applying detection criteria recommended by the Scientific and Standardization Committee of the ISTH on Lupus Anticoagulant. *Thromb. Haemost.* **2013**, *110*, 46–52. [CrossRef] [PubMed]
66. Kaczor, D.A.; Bickford, N.N.; Triplett, D.A. Evaluation of different mixing study reagents and dilution effect in lupus anticoagulant testing. *Am. J. Clin. Pathol.* **1991**, *95*, 408–411. [CrossRef] [PubMed]
67. Kershaw, G.; Orellana, D. Mixing tests: Diagnostic aides in the investigation of prolonged prothrombin times and activated partial thromboplastin times. *Semin. Thromb. Hemost.* **2013**, *39*, 283–290. [PubMed]
68. Moore, G.W.; Culhane, A.P.; Daw, C.R.; Noronha, C.P.; Kumano, O. Mixing test specific cut-off is more sensitive at detecting lupus anticoagulants than index of circulating anticoagulant. *Thromb. Res.* **2016**, *139*, 98–101. [CrossRef] [PubMed]
69. Rosner, E.; Pauzner, R.; Lusky, A.; Modan, M.; Many, A. Detection and quantitative evaluation of lupus circulating anticoagulant activity. *Thromb. Haemost.* **1987**, *57*, 144–147. [PubMed]
70. Favaloro, E.J.; Wong, R.C. Antiphospholipid antibody testing for the antiphospholipid syndrome: A comprehensive practical review including a synopsis of challenges and recent guidelines. *Pathology* **2014**, *46*, 481–495. [CrossRef] [PubMed]
71. Jouhikainen, T. Detection of lupus anticoagulant by means of dilute Russell's viper venom time is affected by oral anticoagulant therapy. *Blood Coagul. Fibrinolysis* **1990**, *1*, 627–632. [PubMed]
72. Olteanu, H.; Downes, K.A.; Patel, J.; Praprotnik, D.; Sarode, R. Warfarin does not interfere with lupus anticoagulant detection by dilute Russell's viper venom time. *Clin. Lab.* **2009**, *55*, 138–142. [PubMed]
73. Chandler, J.B.; Torres, R.; Rinder, H.M.; Tormey, C.A. Lupus anticoagulant testing and anticoagulation do not mix: Quantitation of discrepant results and potential approaches to reduce false positives. *Br. J. Haematol.* **2014**, *167*, 704–707. [CrossRef] [PubMed]

74. Isert, M.; Miesbach, W.; Stoever, G.; Lindhoff-Last, E.; Linnemann, B. Screening for lupus anticoagulants in patients treated with vitamin K antagonists. *Int. J. Lab. Hematol.* **2015**, *37*, 758–765. [CrossRef] [PubMed]

75. Van Os, G.M.; de Laat, B.; Kamphuisen, P.W.; Meijers, J.C.; de Groot, P.G. Detection of lupus anticoagulant in the presence of rivaroxaban using Taipan snake venom time. *J. Thromb. Haemost.* **2011**, *9*, 1657–1659. [CrossRef] [PubMed]

76. Rooney, A.M.; McNally, T.; Mackie, I.J.; Machin, S.J. The Taipan snake venom time: A new test for lupus anticoagulant. *J. Clin. Pathol.* **1994**, *47*, 497–501. [CrossRef] [PubMed]

77. Exner, T. Conceptions and misconceptions in testing for lupus anticoagulants. *J. Autoimmun.* **2000**, *15*, 179–183. [CrossRef] [PubMed]

78. Kitchen, S.; Gray, E.; Mackie, I.; Baglin, T.; Makris, M. BCSH committee. Measurement of non-coumarin anticoagulants and their effects on tests of Haemostasis: Guidance from the British Committee for Standards in Haematology. *Br. J. Haematol.* **2014**, *166*, 830–841. [CrossRef] [PubMed]

79. Thomas, O.; Lybeck, E.; Strandberg, K.; Tynngård, N.; Schött, U. Monitoring low molecular weight heparins at therapeutic levels: Dose-responses of, and correlations and differences between aPTT, anti-factor Xa and thrombin generation assays. *PLoS ONE* **2015**, *10*, e0116835. [CrossRef] [PubMed]

80. Olah, Z.; Szarvas, M.; Bereczky, Z.; Kerenyi, A.; Kappelmayer, J.; Boda, Z. Direct thrombin inhibitors and factor Xa inhibitors can influence the diluted prothrombin time used as the initial screen for lupus anticoagulant. *Arch. Pathol. Lab. Med.* **2013**, *137*, 967–973. [CrossRef] [PubMed]

81. Gosselin, R.; Grant, R.P.; Adcock, D.M. Comparison of the effect of the anti-Xa direct oral anticoagulants apixaban, edoxaban, and rivaroxaban on coagulation assays. *Int. J. Lab. Hematol.* **2016**, *38*, 505–513. [CrossRef] [PubMed]

82. Ratzinger, F.; Lang, M.; Belik, S.; Jilma-Stohlawetz, P.; Schmetterer, K.G.; Haslacher, H.; Perkmann, T.; Quehenberger, P. Lupus-anticoagulant testing at NOAC trough levels. *Thromb. Haemost.* **2016**, *116*, 235–240. [CrossRef] [PubMed]

83. Murer, L.M.; Pirruccello, S.J.; Koepsell, S.A. Rivaroxaban Therapy, False-Positive Lupus Anticoagulant Screening Results, and Confirmatory Assay Results. *Lab. Med.* **2016**, *47*, 275–278. [CrossRef] [PubMed]

84. Arachchillage, D.R.; Mackie, I.J.; Efthymiou, M.; Isenberg, D.A.; Machin, S.J.; Cohen, H. Interactions between rivaroxaban and antiphospholipid antibodies in thrombotic antiphospholipid syndrome. *J. Thromb. Haemost.* **2015**, *13*, 1264–1273. [CrossRef] [PubMed]

85. Merriman, E.; Kaplan, Z.; Butler, J.; Malan, E.; Gan, E.; Tran, H. Rivaroxaban and false positive lupus anticoagulant testing. *Thromb. Haemost.* **2011**, *105*, 385–386. [CrossRef] [PubMed]

86. Hillarp, A.; Gustafsson, K.M.; Faxälv, L.; Strandberg, K.; Baghaei, F.; Fagerberg Blixter, I.; Berndtsson, M.; Lindahl, T.L. Effects of the oral, direct factor Xa inhibitor apixaban on routine coagulation assays and anti-FXa assays. *J. Thromb. Haemost.* **2014**, *12*, 1545–1553. [CrossRef] [PubMed]

87. Góralczyk, T.; Iwaniec, T.; Wypasek, E.; Undas, A. False-positive lupus anticoagulant in patients receiving rivaroxaban: 24 h since the last dose are needed to exclude antiphospholipid syndrome. *Blood Coagul. Fibrinolysis* **2015**, *26*, 473–475. [CrossRef] [PubMed]

88. Moore, G.W.; Peyrafitte, M.; Dunois, C.; Amiral, J. Evaluation of a new formulation dilute Russell's viper venom time for detection of lupus anticoagulants. *J. Thromb. Haemost.* **2016**, *14* (Suppl. 1), 85–86.

89. Sciascia, S.; Breen, K.; Hunt, B.J. Rivaroxaban use in patients with antiphospholipid syndrome and previous venous thromboembolism. *Blood Coagul. Fibrinolysis* **2015**, *26*, 476–477. [PubMed]

90. Kim, Y.A.; Gosselin, R.; Van Cott, E.M. The effects of dabigatran on lupus anticoagulant, diluted plasma thrombin time, and other specialized coagulation assays. *Int. J. Lab. Hematol.* **2015**, *37*, e81–e84. [CrossRef] [PubMed]

91. Bonar, R.; Favaloro, E.J.; Mohammed, S.; Pasalic, L.; Sioufi, J.; Marsden, K. The effect of dabigatran on haemostasis tests: A comprehensive assessment using in vitro and ex vivo samples. *Pathology* **2015**, *47*, 355–364. [CrossRef] [PubMed]

92. Jacquemin, M.; Toelen, J.; Schoeters, J.; van Horenbeeck, I.; Vanlinthout, I.; Debasse, M.; Peetermans, M.; Vanassche, T.; Peerlinck, K.; van Ryn, J.; et al. The addition of idarucizumab to plasma samples containing dabigatran allows the use of routine coagulation assays for the diagnosis of hemostasis disorders. *J. Thromb. Haemost.* **2015**, *13*, 2087–2092. [CrossRef] [PubMed]

93. Devreese, K.; Peerlinck, K.; Hoylaerts, M.F. Thrombotic risk assessment in the antiphospholipid syndrome requires more than the quantification of lupus anticoagulants. *Blood* **2010**, *115*, 870–878. [CrossRef] [PubMed]

94. Schjetlein, R.; Sletnes, K.E.; Wisløff, F. A quantitative, semi-automated and computer-assisted test for lupus anticoagulant. *Thromb. Res.* **1993**, *69*, 239–250. [CrossRef]

95. Jacobsen, E.M.; Barna-Cler, L.; Taylor, J.M.; Triplett, D.A.; Wisløff, F. The Lupus Ratio Test—An interlaboratory study on the detection of lupus anticoagulants by an APTT-based, integrated, and semi-quantitative test. *Thromb. Haemost.* **2000**, *83*, 704–708. [PubMed]

96. Le Querrec, A.; Arnout, J.; Arnoux, D.; Borg, J.Y.; Caron, C.; Darnige, L.; Delahousse, B.; Reber, G.; Sié, P. Quantification of lupus anticoagulants in clinical samples using anti-beta2GP1 and anti-prothrombin monoclonal antibodies. *Thromb. Haemost.* **2001**, *86*, 584–589. [PubMed]

97. De Laat, B.; Wu, X.X.; van Lummel, M.; Derksen, R.H.; de Groot, P.G.; Rand, J.H. Correlation between antiphospholipid antibodies that recognize domain I of beta2-glycoprotein I and a reduction in the anticoagulant activity of annexin A5. *Blood* **2007**, *109*, 1490–1494. [CrossRef] [PubMed]

98. Tripodi, A.; Chantarangkul, V.; Pengo, V. Standardization of lupus anticoagulant. The Lupus Anticoagulant Sensitivity Index (LASI). *Lupus* **2012**, *21*, 715–717. [CrossRef] [PubMed]

99. Kumano, O.; Moore, G.W. Mixing test specific cut-off is more sensitive at detecting in vitro lupus anticoagulant inhibition than the index of circulating anticoagulant with multiple APTT and dRVVT reagents. *J. Thromb. Haemost.* **2016**, *14* (Suppl. 1), 85.

100. Brusch, A. The significance of anti-beta-2-glycoprotein I antibodies in antiphospholipid syndrome. *Antibodies* **2016**, *5*, 16. [CrossRef]

101. Mahler, M.; Albesa, R.; Zohoury, N.; Bertolaccini, M.L.; Ateka-Barrutia, O.; Rodriguez-Garcia, J.L.; Norman, G.L.; Khamashta, M. Autoantibodies to domain 1 of beta 2 glycoprotein I determined using a novel chemiluminescence immunoassay demonstrate association with thrombosis in patients with antiphospholipid syndrome. *Lupus* **2016**, *25*, 911–916. [CrossRef] [PubMed]

![antibodies logo] *antibodies*

MDPI

Review

Neutrophil Extracellular Traps, Antiphospholipid Antibodies and Treatment

Jessica Bravo-Barrera [1,2], Maria Kourilovitch [1,3,*] and Claudio Galarza-Maldonado [1,4]

1   UNERA (Unit of Rheumatic and Autoimmune Diseases), Hospital Monte Sinaí,
    Miguel Cordero 6-111 y av. Solano, Cuenca, Ecuador; jessica.bravo@uneracuenca.com (J.B.-B.);
    claudiogalarza@hotmail.com (C.G.-M.)
2   Department of Hematology and Hemostasis, CDB, Hospital Clinic, Villaroel 170,
    08036 Barcelona, Catalonia, Spain
3   Faculty of Medicine and Health Science, Doctorate Programme "Medicine and Translational Research",
    Barcelona University, Casanova, 143, 08036 Barcelona, Catalonia, Spain
4   Department of Investigation (DIUC-Dirección de Investigación de Universidad de Cuenca),
    Cuenca State University, Av. 12 de Abril y Agustin Cueva, Cuenca, Ecuador
*   Correspondence: maria.kourilovitch@uneracuenca.com; Tel.: +593-99-353-8563

Academic Editor: Ricard Cervera
Received: 1 January 2017; Accepted: 1 March 2017; Published: 6 March 2017

Abstract: Neutrophil extracellular traps (NETs) are a network of extracellular fibers, compounds of chromatin, neutrophil DNA and histones, which are covered with antimicrobial enzymes with granular components. Autophagy and the production of reactive oxygen species (ROS) by nicotinamide adenine dinucleotide phosphate (NADPH) oxidase are essential in the formation of NETs. There is increasing evidence that suggests that autoantibodies against beta-2-glycoprotein-1 (B2GP1) induce NETs and enhance thrombosis. Past research on new mechanisms of thrombosis formation in antiphospholipid syndrome (APS) has elucidated the pharmacokinetics of the most common medication in the treatment of the disease.

Keywords: neutrophil extracellular traps; NETosis; autophagy; antibodies; antiphospholipid syndrome

1. Introduction

Neutrophils are granulocytes that have an essential role in the pathology of a broad spectrum of inflammatory diseases. In circulation, the neutrophils remain inactive; but under inflammatory conditions, they are recruited to the tissues, where they participate in the destruction of pathogens through different mechanisms. The neutrophils' activation occurs via a variety of receptors, including pattern-recognition receptors and Fc-receptors [1]. For decades, phagocytosis was considered the primary mechanism by which neutrophils targeted infections [2]. However, in 2004, Brinkmann et al. described another distinct antimicrobial activity of neutrophils, in which neutrophils were shown to release extracellular traps (NETs) [3]. Steinberg and Grinstein named this process of neutrophil cell death as "NETosis" [4].

NETs are a network of extracellular fibers, compounds of decondensed chromatin, including neutrophil DNA and high affinity histones, which are covered with antimicrobial enzymes and granular components, such as myeloperoxidase (MPO), neutrophil elastase (NE), cathepsin G and other microbicidal peptides [3,5].

In vitro studies, using the non-physiological stimulus phorbol-12-myristate-13-acetate (PMA), demonstrated that during NETs formation, a rupture of the cell membrane and exposure of the inner membrane phospholipids occur. NETosis was classified as a novel type of cell death [6]. However, there is an ongoing controversy on whether or not the death of neutrophils actually occurs in vivo.

Through detailed observations of neutrophil behavior on Gram-positive skin infections in mice and humans, Yipp et al. were able to demonstrate that while neutrophils form and release NETs during crawling and become anuclear, they do not show any signs of programmed cell death [7].

Further studies are needed to elucidate whether or not anuclear neutrophils have the capacity to activate other cell mechanisms and functions [8].

The interest in the role of NETs in autoimmune diseases arose with the discovery of certain mechanisms that trigger NETosis by non-infectious stimuli, such as: immune complexes, autoantibodies, cytokines, cholesterol and monosodium urate (MSU) crystals [1]. Multiple studies have shown the implication of such mechanisms in NETs formation in chronic inflammatory processes, as seen in lung [9], systemic lupus erythematosus [10], antineutrophil cytoplasmic antibodies (ANCA)-associated vasculitis [11], rheumatoid arthritis [12], gouty arthritis [13,14], familiar Mediterranean fever [15], psoriasis [16] and autoimmune coagulation disorders [17,18].

In susceptible individuals, many of the molecules released through NETosis (for example, double-stranded (ds) DNA, histones, cytokines, MPO, etc.) could be recognized by the immune system as autoantigens and initiate the autoimmune response. If this occurs, a vicious cycle of autoimmune reactions is triggered, which leads to further release of antigenic material [19].

In this review, we will address the contribution of NETosis in the development of antiphospholipid-mediated pathology. Furthermore, we will identify NETosis-related aspects of the pharmacokinetics of medication used in the treatment of APS.

2. NETs Formation

During NETs formation, the neutrophils lose their variability, which results in the activation of certain signaling pathways producing the dissolution of the nuclear envelope [6]. Remijsen, et al. proved that autophagy and the production of reactive oxygen species (ROS) by NADPH-oxidase are essential in the formation of NETs [20]. The NADPH enzyme is activated in response to the threat of infection, triggering the generation of antimicrobial reactive oxidants [21]. The inhibition of either autophagy or NADPH-oxidase prevents decondensation of intracellular chromatin; without the ability to complete these processes, NETosis cannot occur [20,22].

ROS is a signaling molecule that can promote inflammation and tissue damage [23]. The generation of ROS is necessary for the activation of neutrophil enzymes, which produce DNA unwinding, a critical process in NETosis [24]. As NETosis is dependent on ROS production by NADPH-oxidase, the inability to form ROS in genetically-defective NADPH-oxidase patients prevents NETs formation [6,25].

Cytokines are activators of neutrophil functions and, consequently, play an important role in the process of NETosis. The neutrophils of healthy subjects, treated with TNF-α, IL-1β or IL-8, produce free radicals, and NETs form by the activation of NADPH-oxidase. This findings point out the importance of cytokines in the enhanced release of NETs in systemic inflammatory responses syndrome [26]. Cytokines, such as TNF-α, IL-1β, IL-8 and IL-6, have been observed to enhance free radical generation. Moreover, a variety of studies emphasize the significant role of TNF-α in mitochondrial ROS production [27,28].

It is important to note that aggregated NETs have been observed to regulate inflammation through the degradation of cytokines and chemokines, limiting the inflammation in patients with MSU deposits [29].

Platelets are one of the important actors in the immune response and play a critical role in NETs formation [30]. When platelets stimulation occurs, they begin to secrete molecules that can modulate the activation of neutrophils. One such molecule is high mobility group box 1 (HMGB1), a damage-associated molecular pattern molecule. HMGB1 is released as a result of cell death and is an important marker of inflammatory response to tissue damage. Recently, it has been demonstrated that the HMGB1-platelets complex is one of the key inductors of NETs formation. In addition, HMGB1 regulates cell death through the management of apoptosis, autophagy and necrosis in cells [31,32].

The capacity of HMGB1 to inhibit apoptosis can explain the absence of observed cell death in anuclear neutrophils following NETs in vivo.

2.1. Autophagy and NETosis

Autophagy was defined over 40 years ago by Christian de Duve as the "eating of self" [33], and through the work of Yoshinori Ohsumi (2016 Nobel Prize winner in physiology or medicine), the mechanisms and genes of autophagy have been elucidated [34,35].

Autophagy is an important mechanism for the preservation of cell integrity and survival. By recycling cytosolic macromolecules and organelles, autophagy provides essential nutrients and the clearance of cellular proteins [20,36]. In recent years, the role of autophagy has been discussed in relation to a spectrum of diseases, such as cancer, neurodegenerative, autoimmune and cardiovascular diseases [37].

Autophagy occurs in the nucleated cells of an organism. The process of autophagy in platelets is an important regulator of intra-vascular NETs formation and thrombosis [17]. Ouseph, et al. demonstrated that the process of autophagy not only occurs when platelets are at rest, but also during their activation. A deficient autophagy can produce unidentified platelet dysfunction [38].

In regard to autoimmune processes, the function of autophagy as a promotor of the survival of cells resistant to apoptosis is a current topic of investigation. Amaravadi et al. postulate that autophagy can be an adaptive mechanism that contributes to cell survival and resistance to therapy-induced apoptosis in a Myc-induced model of lymphoma [39]. Likewise, disbalance in immunologic-related function, such as the removal of intracellular pathogens, secretory pathways (including vesicle trafficking), autophagic regulation of ROS, pro-inflammatory signaling and antigen presentation, often trigger autoimmunity [40].

Cytokines play an important role in the regulation of autophagy. The processing and secretion of IL-1b, IL-18 and IL-1a by macrophages and dendritic cells are negatively regulated by autophagy. Conversely, autophagy positively regulates the transcription and secretion of TNF-α, IL-8 and, possibly, IL-6 and type I IFN [41]. Toll-like receptors (TLR) and NOD-like receptors (NLR) are potent inducers of autophagy due to their ability to recognize different pathogens, stress factors and cytokines [40,42].

2.2. NETs in Antiphospholipid Syndrome and Thrombosis

Antiphospholipid syndrome (APS) is an autoimmune disease characterized by the presence of elevated titers of antiphospholipid antibodies (aPL). These antibodies are predisposed to arterial and venous thrombosis and fetal loss [43].

One of the dominating autoantibodies in this syndrome targets beta-2-glycoprotein 1 (B2GP1), a circulating phospholipid-binding glycoprotein, secreted by the liver, monocytes, trophoblasts, endothelial cells and platelets [44]. The presence of anti-B2GP1 is frequently associated with thrombotic events, pro-atherogenic mechanisms and vascular cell dysfunction [45].

The definition of APS, according to the Sidney Classification Criteria, states that there must be clinical evidence of vascular thrombosis and/or pregnancy-related morbidity and one of the following laboratory criteria: anticardiolipin antibodies, anti-B2GP1 antibodies or lupus anticoagulant. Furthermore, in order to be classified as APS, there should be at least 12 weeks, and no more than five years, between the clinical manifestation and the positive aPL test [46].

Actually, there is no targeted treatment for APS, and current therapies focus on the management of thrombosis with long-term anticoagulant medication [47]. The mechanisms by which antiphospholipid antibodies induce thrombosis are still unclear.

Neutrophils have been observed to be significantly related to arterial and venous thrombosis. During the autoimmune process, NETs components can be recognized by the immune system as an autoantigen that directly or indirectly influence the pathogenesis of a variety of inflammatory and autoimmune diseases.

In recent years, studies on NETs have revealed evidence that autoantibodies against B2GP1 induce NETs and enhance thrombosis. Yalavarthi, et al. [48] described the release of NETs, promoted by anticardiolipin antibodies, as a new possible mechanism of thrombosis in antiphospholipid syndrome. Confirming the hypothesis that antiphospholipid antibodies activate neutrophils to release NETs, the investigators demonstrated that isolated neutrophils of the patients with APS enhanced spontaneous NETs release, when compared with controls. In addition, a positive correlation between anti-B2GP1 IgG, lupus anticoagulant, anticardiolipin IgG and circulating MPO-DNA complexes was found, showing a correlation between the level of circulating MPO-DNA complexes and NETs in vivo. However, no correlation was observed between MPO-DNA and anti-cardiolipin antibodies IgM and IgA. A significant statistical difference was confirmed between "triple-positive" patients for lupus anticoagulant, anti-B2GP1 IgG and anti-cardiolipin IgG antibodies and "single-positive" patients and their subsequent correlation with MPO-DNA levels. The stimulation of neutrophils with isolated total IgG fractions from "triple-positive" patients with APS produces significant NETs release when compared with healthy controls. After the depletion of the anti-B2GP1 IgG fraction, the NETs abrogate. By utilizing different laboratory methods, B2GP1 was detected on the neutrophils' surface. This discovery can explain the binding of anti-B2GP1 antibodies with neutrophils and the consequent triggering of NETosis. Another interesting observation was that both ROS formation and TLR4 engagement were required for aPL-mediated NETs release. In contrast, PMA-stimulated NETosis was TLR4-independent. These data enable one to consider the TLR4 as a possible mediator of aPL stimulation in neutrophils.

In a recently published study, Meng et al. demonstrated, through mice models in vivo, that the administration of IgG in APS patients had a prothrombotic effect. Moreover, APS thrombi were enriched in NETs. Thus, the stimulation of mouse neutrophils by APS IgG resulted in NETosis. In addition, this group of researchers showed that both neutrophil depletion and DNase administration have been seen to abrogate thrombosis in APS mice [49].

While aPL/neutrophil interplay in obstetric APS is still unknown and further investigation is required, a number of studies suggest a pathogenic role of NETs in aPL-negative patients experiencing pre-eclampsia [50].

Leffler et al. proved that patients with systemic lupus erythematosus (SLE) have a defect in DNase-mediated NETs degradation [51]. Nevertheless, this phenomenon is not significant in patients with APS; and if present, does not correlate with the presence of aPL antibodies, such as anti-B2GP1, anti-cardiolipins or lupus anticoagulant. There is no evidence that aPL antibodies coincide with or cause failed NETs degradation [52].

NETs contribute both to arterial and venous thrombosis through the following mechanisms: its ability to bind and activate platelets, tissue factor (TF) and coagulation factor VII, which accelerate the thrombus formation [38].

Kambas et al. focused on the role of neutrophils in the coordination between inflammation and coagulation. The researchers demonstrated that TF-bearing NETs released from the neutrophils of patients with sepsis play a key role in the activation of the coagulation system by triggering thrombin generation. Furthermore, it was shown that the autophagy-dependent mechanism is involved in the extracellular localization of TF in NETs [53,54]. In another study, this group of investigators propose that TF expressed by NETs, as well as the TF expressed by microparticles could be the trigger of a new mechanism for the induction of inflammation and thrombosis in active ANCA-associated vasculitis [55].

In vitro and in vivo studies have shown that NETs contribute to thrombus formation and coagulation factors involved in clotting [56,57] through a variety of components: high amounts of TF expressed by NETs at sites of inflammation produce localized activation of the coagulation cascade; the DNA component of NETs activates factor XII, initiating contact pathway coagulation, leading to fibrin formation; histones, components of extracellular nucleosomes in NETs, activate platelets and sequester certain anticoagulant molecules like thrombomodulin and protein C. In addition, neutrophil serine proteases (neutrophil elastase and cathepsin G), present in NETs, generate degradation and inactivation of the anticoagulant molecule tissue factor pathway inhibitor (TFPI). Finally, NETs suppress fibrinolysis by intercalating into the fibrin clots [5,18,22,58–60].

Additional information on how these mechanisms secure the release of NETs is necessary in order to better understand the physiological conditions of neutrophils' function. The unique link between inflammation and thrombosis is extracellular DNA. When tested, it was discovered that markers of extracellular DNA traps are abundant in deep venous thrombosis (DVT) [5].

Maternal TF on neutrophils is a necessary trigger in the pathogenesis of APS, which results in fetal loss. This demonstrates an important connection between complement components, TF and neutrophils [61].

The significant role of TF in thrombosis is based on vascular injury by factor VIIa binding. Furthermore, it has been established that TF is important in thrombosis and inflammation in APS patients [62]. Ritis, et al. observed that the neutrophils of healthy individuals stimulated with APS serum are able to express TF [63]. Moreover, the interaction of complement with neutrophils produces the generation of TF-dependent coagulation activity and the induction of TF-dependent thrombosis. This interaction occurs through C5a, a potent chemotactic factor, which is activated through C5aR receptors expressed on their surfaces. After activation, neutrophils migrate to inflamed tissues, infiltrating the injured sites [61].

Increasing evidence shows that neutrophils are related to obstetric antiphospholipid syndrome, in which pathogenic NETosis is initiated by aPL binding trophoblasts. This binding produces the activation of complement cascade leading to C5a generation. The involvement of C5a with a C5a receptor on neutrophils produces the TF expression. The TF expression increases cellular activation (ROS production), leading to inflammation, injury and fetal death [64]. (See Figure 1).

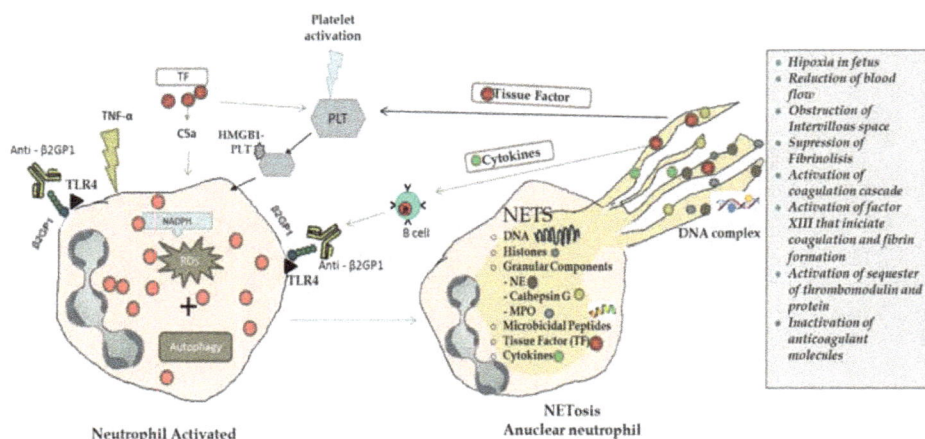

Figure 1. Trigger factors, such as activated platelets through the HMGB1-platelet complex, pro-inflammatory cytokines (TNF-α), tissue factor (TF) or the interaction of anti-B2GP1 with surface B2GP1 in antiphospholipid syndrome via TLR4, prompt NETosis. ROS by NADPH and autophagy induce NETs formation process. During NETosis, the components of NETs (DNA complex, histones, microbicidal peptides, cytokines, granular components) are released. If this autoimmune vicious cycle occurs, TF produced by NETs also activates platelets, as well as cytokines from NETs participation in the activation of B cells to produce autoantibodies. ROS: reactive oxygen species; NADPH: nicotinamide adenine dinucleotide phosphate; PLT: platelets; HMGB1: high mobility group box 1; TNF-α: tumor necrosis factor α; Anti-B2GP1: anti-beta 2 glycoprotein 1; B2GP1: beta 2 glycoprotein 1; TLR4: Toll-like receptor 4; NE: neutrophil elastase; MPO: myeloperoxidase; DNA: deoxyribonucleic acid; TF: tissue factor.

3. New Mechanisms of Old Therapeutics Agents

Evidence of NETs formation and its relationship with thrombosis has led to the increased investigation of new mechanisms of action and the existent drugs.

3.1. Acetylsalicylic Acid

Acetylsalicylic acid (ASA) in low dose has been widely used as a therapy for obstetric APS due to its antiplatelet mechanism of action by the inhibition of platelet cyclooxygenase [65]. Lapponi et al. demonstrated that ASA and nuclear factor NF-kB inhibitors significantly decrease NETs generation from neutrophils stimulated with phorbol 12-myristate 13-acetate (PMA) or TNF-α; while dexamethasone has no such effect [66].

3.2. Heparins

Heparins, a mixture of multifunction glycosaminoglycans, are principal drugs in the treatment of thrombosis and thromboprophylaxis in high-risk patients with obstetric APS. These drugs have both antithrombin-dependent and antithrombin independent activities. Heparins have the ability to almost completely dismantle NETs through the destabilization of backbone formed by chromatin fibers. In addition, they remove platelet aggregations and releases histones from chromatin, interfering with neutrophil-platelet cross-talk [67].

The capacity of heparin to block the binding of HMGB1 to the surface of macrophages also contributes in the control of NETosis through inhibiting the induction of pro-inflammatory cytokines, including TNF-α [68].

Moreover, pre-treatment with low-molecular-weight heparins (LMWH) has an effect on the induction of autophagy and NETs formation in vitro and in vivo: LMWHs at a "prophylactic dose", used for the prevention of obstetric complications related to APS, inhibit the ability of neutrophils to activate autophagy, to mobilize the granule content and to form NETs [69].

3.3. Hydroxychloroquine and Chloroquine

Hydroxychloroquine (HCQ) and chloroquine (CQ) are antimalarial immunomodulators. The antimalarials are a cornerstone in the treatment of SLE and APS. HCQ has been shown to reduce the risk of thromboembolic events in both patients with SLE and positive aPL. These drugs block the processing of NETs through TLR9 in plasmacytoid dendritic cells (pDCs) [70].

CQ significantly inhibits NETs formation in controls and lupus nephritis neutrophils in vitro [71]. CQ also plays an important role in regulating NETosis through its autophagy inhibitor property. CQ has an effect on the lysosomal degradation pathway, enhancing the autophagic vesicle clearance. HCQ, a derivative of CQ, has a similar mechanism of action [39,72].

3.4. Vitamin D

The immunomodulator vitamin D calcitriol/1,25(OH)2D3 reduces the production of the mediators of the inflammation and ROS in neutrophils [73]. Handono, et al. found that vitamin D calcitriol/1,25(OH)2D3 could decrease NETosis activity and reduce endothelial damage in patients with SLE and hypovitaminosis D [74].

3.5. Vitamin C

Vitamin C, as an endogenous antioxidant, is essential in diseases prevention. It was discovered that vitamin C operates as a novel regulator of NETs formation in pathways associated with sepsis. An increase of vitamin C has been shown to weaken NETosis in septic mice. Furthermore, polymorphonuclear cells, deficient in vitamin C, were more susceptible to produce NETs via NFκB activation, which develop ROS production and autophagy, indispensable factors for NETs formation [75].

Vitamin C-deficient neutrophils show an increase of the expression of peptidyl arginine deiminase 4 (PAD4). Furthermore, citrullination with PAD4 produces chromatin decondensation, which is essential in NETs formation [75,76].

Other evidence confirms that vitamin C attenuates NETosis induced by PMA in neutrophils from healthy volunteers [75].

However, various randomized studies could not demonstrate the effectiveness of vitamin C supplementation in preventing cardiovascular events, including stroke [77–79].

4. Biologic Anti-Cytokine Therapy

As previously discussed, cytokines play an important role in the process of NETosis. It was demonstrated that the inhibition of TNF and IL-17 abates NETosis in patients with rheumatoid arthritis [80]. Several authors find the administration of TNF-α inhibitors (adalimumab, etanercept, infliximab) useful in the treatment of refractory obstetric antiphospholipid syndrome [81]. Nevertheless, special attention must be paid to certolizumab pegol, the PEGylated Fab' fragment of humanized anti-TNF-α monoclonal antibody, as a potential treatment of this condition [82]; due to the fact that this TNF-α inhibitor has a minimal placental transfer, measured by cord blood levels at birth, when compared with infliximab and adalimumab [83].

4.1. Statins

The pleiotropic immunomodulatory effect, anti-inflammatory and anti-thrombotic properties of statins have interested researchers and physicians during the last few decades [84]. The ability of statins to downregulate tissue factor and other prothrombotic markers was described by several researchers [85–87]. Nevertheless, regarding NETosis, Chow et al. [88] demonstrated that statins enhance NETs production despite the existing evidence of its capability to reduce ROS production [89]. The results of this study suggest that statins can promote NETs formation in response to a lower threshold level of ROS signaling.

Thus, although the boost of NETosis by statins has been shown to be useful in the treatment of sepsis and other infectious diseases, which lead to immunosuppression [90], the same effect can explain incidences of statin-related autoimmune reactions [91–93].

4.2. Potential Therapeutic Agents

The possibility to modulate NETosis demands more research on new therapeutic opportunities. Among molecules that have potential effect on neutrophil and NETs formation are: DNase 1 (enzymatic degradation of NETs) [52,53], eculizumab (anti-C5a monoclonal antibody, reduce neutrophil activation) [94], rituximab and belimumab (B cell depletion, downregulation of NETs formation through control of antibodies production) and Resatrovid (TAK-242, a small-molecule-specific inhibitor of Toll-like receptor 4 signaling, inhibitor of NETs release by human neutrophils) [19,95].

5. Conclusions

There is much evidence with respect to the participation of NETs in thrombotic events. Nevertheless, more investigation is needed to completely elucidate the role of the aPL in NETs formation, as well as its participation in the pathologic mechanisms of the APS, especially obstetric APS. Mechanisms that involve NETs in pathologic processes may differ in vivo and in vitro. Furthermore, the structure and property of NETs might vary depending on the pathological and physiological conditions. Continued research on the mechanisms of action of current market drugs, as well as the advancing development of new medication, will evolve treatments for patients diagnosed with different forms of APS.

Acknowledgments: We thank Sarah Anne Bradley for her work of translation and English revision of the paper

Author Contributions: All authors designed the review and wrote the manuscript.

Conflicts of Interest: The authors declare no conflict of interest.

References

1. Grayson, P.C.; Schauer, C.; Herrmann, M.; Kaplan, M.J. Review: Neutrophils as Invigorated Targets in Rheumatic Diseases. *Arthritis Rheumatol.* **2016**, *68*, 2071–2082. [CrossRef] [PubMed]
2. Kaplan, J.M. Neutrophil extracelullar traps (NETs): Double-edged swords of innate immunity 1. *J. Immunol.* **2013**, *189*, 2689–2695. [CrossRef] [PubMed]
3. Brinkmann, V. Neutrophil Extracellular Traps Kill Bacteria. *Science* **2004**, *303*, 1532–1535. [CrossRef] [PubMed]
4. Steinberg, B.E.; Grinstein, S. Unconventional roles of the NADPH oxidase: Signaling, ion homeostasis, and cell death. *Sci. STKE* **2007**, *2007*, pe11. [CrossRef] [PubMed]
5. Gould, T.J.; Lysov, Z.; Liaw, P.C. Extracellular DNA and histones: Double-edged swords in immunothrombosis. *J. Thromb. Haemost.* **2015**, *13*, 82–91. [CrossRef] [PubMed]
6. Fuchs, T.A.; Abed, U.; Goosmann, C.; Hurwitz, R.; Schulze, I.; Wahn, V.; Weinrauch, Y.; Brinkmann, V.; Zychlinsky, A. Novel cell death program leads to neutrophil extracellular traps. *J. Cell Biol.* **2007**, *176*, 231–241. [CrossRef] [PubMed]
7. Yipp, B.G.; Petri, B.; Salina, D.; Jenne, C.N.; Scott, B.N.; Zbytnuik, L.D.; Pittman, K.; Asaduzzaman, M.; Wu, K.; et al. Infection-induced NETosis is a dynamic process involving neutrophil multitasking in vivo. *Nat. Med.* **2012**, *18*, 1386–1393. [CrossRef] [PubMed]
8. Darrah, E.; Andrade, F. NETs: The missing link between cell death and systemic autoimmune diseases? *Front. Immunol.* **2013**, *3*, 428. [CrossRef] [PubMed]
9. Porto, B.N.; Stein, R.T. Neutrophil Extracellular Traps in Pulmonary Diseases: Too Much of a Good Thing? *Front. Immunol.* **2016**, *7*, 311. [CrossRef] [PubMed]
10. Boilard, E.; Fortin, P.R. Connective tissue diseases: Mitochondria drive NETosis and inflammation in SLE. *Nat. Rev. Rheumatol.* **2016**, *12*, 195–196. [CrossRef] [PubMed]
11. Söderberg, D.; Segelmark, M. Neutrophil Extracellular Traps in ANCA-Associated Vasculitis. *Front. Immunol.* **2016**, *7*, 256. [CrossRef] [PubMed]
12. Corsiero, E.; Pratesi, F.; Prediletto, E.; Bombardieri, M.; Migliorini, P. NETosis as Source of Autoantigens in Rheumatoid Arthritis. *Front. Immunol.* **2016**, *7*, 485. [CrossRef] [PubMed]
13. Pisetsky, D.S. Gout, tophi and the wonders of NETs. *Arthritis Res. Ther.* **2014**, *16*, 431. [CrossRef] [PubMed]
14. Mitroulis, I.; Kambas, K.; Chrysanthopoulou, A.; Skendros, P.; Apostolidou, E.; Kourtzelis, I.; Drosos, G.; Boumpas, D.; Ritis, K. Neutrophil Extracellular Trap Formation Is Associated with IL-1β and Autophagy-Related Signaling in Gout. *PLoS ONE* **2011**, *6*, e29318. [CrossRef] [PubMed]
15. Apostolidou, E.; Skendros, P.; Kambas, K.; Mitroulis, I.; Konstantinidis, T.; Chrysanthopoulou, A.; Nakos, K.; Tsironidou, V.; Koffa, M.; Boumpas, D.T.; Ritis, K. Neutrophil extracellular traps regulate IL-1β-mediated inflammation in familial Mediterranean fever. *Ann. Rheum. Dis.* **2014**, 1–9. [CrossRef] [PubMed]
16. Lin, A.M.; Rubin, C.J.; Khandpur, R.; Wang, J.Y.; Riblett, M.; Yalavarthi, S.; Villanueva, E.C.; Shah, P.; Kaplan, M.J.; Bruce, A.T. Mast cells and neutrophils release IL-17 through extracellular trap formation in psoriasis. *J. Immunol.* **2011**, *187*, 490–500. [CrossRef] [PubMed]
17. Kimball, A.S.; Obi, A.T.; Diaz, J.A.; Henke, P.K. The emerging role of NETs in venousthrombosis and immunothrombosis. *Front. Immunol.* **2016**, *7*, 1–8. [CrossRef] [PubMed]
18. Rao, A.N.; Kazzaz, N.M.; Knight, J.S. Do neutrophil extracellular traps contribute to the heightened risk of thrombosis in inflammatory diseases? *World J. Cardiol.* **2015**, *7*, 829–842. [CrossRef] [PubMed]
19. Gupta, S.; Kaplan, M.J. The role of neutrophils and NETosis in autoimmune and renal diseases. *Nat. Rev. Nephrol.* **2016**, *12*, 402–413. [CrossRef] [PubMed]
20. Remijsen, Q.; Vanden Berghe, T.; Wirawan, E.; Asselbergh, B.; Parthoens, E.; De Rycke, R.; Noppen, S.; Delforge, M.; Willems, J.; Vandenabeele, P. Neutrophil extracellular trap cell death requires both autophagy and superoxide generation. *Cell Res.* **2011**, *21*, 290–304. [CrossRef] [PubMed]
21. Almyroudis, N.G.; Grimm, M.J.; Davidson, B.A.; Röhm, M.; Urban, C.F.; Segal, B.H. NETosis and NADPH oxidase: At the intersection of host defense, inflammation, and injury. *Front. Immunol.* **2013**, *4*, 45. [CrossRef] [PubMed]
22. Iba, T.; Hashiguchi, N.; Nagaoka, I.; Tabe, Y.; Murai, M. Neutrophil cell death in response to infection and its relation to coagulation. *J. Intensive Care* **2013**, *1*, 13. [CrossRef] [PubMed]

23. Glick, D.; Barth, S.; Macleod, K.F. Autophagy: Cellular and molecular mechanisms. *J. Pathol.* **2010**, *221*, 3–12. [CrossRef] [PubMed]
24. Craft, J.E. Dissecting the immune Cell Mayhem that Drives Lupus Pathogenesis. *Sci. Transl. Med.* **2011**, *3*, 9–16. [CrossRef] [PubMed]
25. Björnsdottir, H.; Welin, A.; Michaëlsson, E.; Osla, V.; Berg, S.; Christenson, K.; Sundqvist, M.; Dahlgren, C.; Karlsson, A.; Bylund, J. Neutrophil NET formation is regulated from the inside by myeloperoxidase-processed reactive oxygen species. *Free Radic. Biol. Med.* **2015**, *89*, 1024–1035. [CrossRef] [PubMed]
26. Keshari, R.S.; Jyoti, A.; Dubey, M.; Kothari, N.; Kohli, M.; Bogra, J.; Barthwal, M.K.; Dikshit, M. Cytokines Induced Neutrophil Extracellular Traps Formation: Implication for the Inflammatory Disease Condition. *PLoS ONE* **2012**, *7*, e48111. [CrossRef] [PubMed]
27. Baregamian, N.; Song, J.; Bailey, C.E.; Papaconstantinou, J.; Evers, B.M.; Chung, D.H. Tumor necrosis factor-alpha and apoptosis signal-regulating kinase 1 control reactive oxygen species release, mitochondrial autophagy, and c-Jun N-terminal kinase/p38 phosphorylation during necrotizing enterocolitis. *Oxid. Med. Cell. Longev.* **2009**, *2*, 297–306. [CrossRef]
28. Xue, X.; Piao, J.-H.; Nakajima, A.; Sakon-Komazawa, S.; Kojima, Y.; Mori, K.; Yagita, H.; Okumura, K.; Harding, H.; Nakano, H. Tumor necrosis factor alpha (TNFalpha) induces the unfolded protein response (UPR) in a reactive oxygen species (ROS)-dependent fashion, and the UPR counteracts ROS accumulation by TNFalpha. *J. Biol. Chem.* **2005**, *280*, 33917–33925. [CrossRef] [PubMed]
29. Schauer, C.; Janko, C.; Munoz, L.E.; Zhao, Y.; Kienhöfer, D.; Frey, B.; Lell, M.; Manger, B.; Rech, J.; Naschberger, E.; et al. Aggregated neutrophil extracellular traps limit inflammation by degrading cytokines and chemokines. *Nat. Med.* **2014**, *20*, 511–517. [CrossRef] [PubMed]
30. Carestia, A.; Kaufman, T.; Schattner, M. Platelets: New Bricks in the Building of Neutrophil Extracellular Traps. *Front. Immunol.* **2016**, *7*, 271. [CrossRef] [PubMed]
31. Maugeri, N.; Campana, L.; Gavina, M.; Covino, C.; De Metrio, M.; Panciroli, C.; Maiuri, L.; Maseri, A.; D'Angelo, A.; Bianchi, M.E.; et al. Activated platelets present high mobility group box 1 to neutrophils, inducing autophagy and promoting the extrusion of neutrophil extracellular traps. *J. Thromb. Haemost.* **2014**, *12*, 2074–2088. [CrossRef] [PubMed]
32. Stark, K.; Philippi, V.; Stockhausen, S.; Busse, J.; Antonelli, A.; Miller, M.; Schubert, I.; Hoseinpour, P.; Chandraratne, S.; von Brühl, M.L.; et al. Disulfide HMGB1 derived from platelets coordinates venous thrombosis in mice. *Blood* **2016**, *128*, 2435–2449. [CrossRef]
33. Deter, R.L.; De Duve, C. Influence of glucagon, an inducer of cellular autophagy, on some physical properties of rat liver lysosomes. *J. Cell Biol.* **1967**, *33*, 437–449. [CrossRef] [PubMed]
34. Ohsumi, Y. Yoshinori Ohsumi: Autophagy from beginning to end. Interview by Caitlin Sedwick. *J. Cell Biol.* **2012**, *197*, 164–165. [CrossRef] [PubMed]
35. Nobel Prize Honors Autophagy Discovery. *Cancer Discov.* **2016**, *6*, 1298–1299. [CrossRef]
36. Sha, L.-L.; Wang, H.; Wang, C.; Peng, H.-Y.; Chen, M.; Zhao, M.-H. Autophagy is induced by anti-neutrophil cytoplasmic Abs and promotes neutrophil extracellular traps formation. *Innate Immun.* **2016**, *22*, 658–665. [CrossRef] [PubMed]
37. Ryter, S.W.; Lee, S.-J.; Smith, A.; Choi, A.M.K. Autophagy in Vascular Disease. *Proc. Am. Thorac. Soc.* **2010**, *7*, 40–47. [CrossRef] [PubMed]
38. Ouseph, M.M.; Huang, Y.; Banerjee, M.; Joshi, S.; MacDonald, L.; Zhong, Y.; Liu, H.; Li, X.; Xiang, B.; Zhang, G.; et al. Autophagy is induced upon platelet activation and is essential for hemostasis and thrombosis. *Blood* **2015**, *126*, 1224–1233. [CrossRef] [PubMed]
39. Amaravadi, R.K.; Yu, D.; Lum, J.J.; Bui, T.; Christophorou, M.A.; Evan, G.I.; Thomas-Tikhonenko, A.; Thompson, C.B. Autophagy inhibition enhances therapy-induced apoptosis in a Myc-induced model of lymphoma. *J. Clin. Investig.* **2007**, *117*, 326–336. [CrossRef] [PubMed]
40. Levine, B.; Mizushima, N.; Virgin, H.W. Autophagy in immunity and inflammation. *Nature* **2011**, *469*, 323–335. [CrossRef] [PubMed]
41. Harris, J. Autophagy and cytokines. *Cytokine* **2011**, *56*, 140–144. [CrossRef] [PubMed]
42. Brinkmann, V.; Zychlinsky, A. Neutrophil extracellular traps: Is immunity the second function of chromatin? *J. Cell Biol.* **2012**, *198*, 773–783. [CrossRef] [PubMed]

43. Rottem, M.; Krause, I.; Fraser, A.; Stojanovich, L.; Rovensky, J.; Shoenfeld, Y. Antiphospholipid syndrome. *Lupus* **2006**, *95*, 336–342. [CrossRef]

44. Conti, F.; Sorice, M.; Circella, A.; Alessandri, C.; Pittoni, V.; Caronti, B.; Calderaro, C.; Griggi, T.; Misasi, R.; Valesini, G. Beta-2-glycoprotein I expression on monocytes is increased in anti-phospholipid antibody syndrome and correlates with tissue factor expression. *Clin. Exp. Immunol.* **2003**, *132*, 509–516. [CrossRef] [PubMed]

45. Sorice, M.; Buttari, B.; Capozzi, A.; Profumo, E.; Facchiano, F.; Truglia, S.; Recalchi, S.; Alessandri, C.; Conti, F.; Misasi, R.; et al. Antibodies to age-β$_2$glycoprotein I in patients with anti-phospholipid antibody syndrome. *Clin. Exp. Immunol.* **2016**, *184*, 174–182. [CrossRef] [PubMed]

46. Miyakis, S.; Lockshin, M.D.; Atsumi, T.; Branch, D.W.; Brey, R.L.; Cervera, R.; Derksen, R.H.; DE Groot, P.G.; Koike, T.; Meroni, P.L.; et al. International consensus statement on an update of the classification criteria for definite antiphospholipid syndrome (APS). *J. Thromb. Haemost.* **2006**, *4*, 295–306. [CrossRef] [PubMed]

47. Chaturvedi, S.; Mccrae, K.R. The antiphospholipid syndrome: still an enigma. *ASH Educ. Progr. B.* **2015**, 53–60. [CrossRef] [PubMed]

48. Yalavarthi, S.; Gould, T.J.; Rao, A.N.; Mazza, L.F.; Morris, A.E.; Núñez-Álvarez, C.; Hernández-Ramírez, D.; Bockenstedt, P.L.; Liaw, P.C.; Cabral, A.R.; et al. Release of neutrophil extracellular traps by neutrophils stimulated with antiphospholipid antibodies: A newly identified mechanism of thrombosis in the antiphospholipid syndrome. *Arthritis Rheumatol.* **2015**, *67*, 2990–3003. [CrossRef] [PubMed]

49. Meng, H.; Yalavarthi, S.; Kanthi, Y.; Mazza, L.F.; Elfline, M.A.; Luke, C.E.; Pinsky, D.J.; Henke, P.K.; Knight, J.S.; et al. In vivo role of neutrophil extracellular traps in antiphospholipid antibody-mediated venous thrombosis. *Arthritis Rheumatol.* **2017**, *69*, 655–667. [CrossRef] [PubMed]

50. Hahn, S.; Giaglis, S.; Hoesli, I.; Hasler, P. Neutrophil NETs in reproduction: From infertility to preeclampsia and the possibility of fetal loss. *Front. Immunol.* **2012**, *3*, 362. [CrossRef]

51. Leffler, J.; Gullstrand, B.; Jönsen, A.; Nilsson, J.Å.; Martin, M.; Blom, A.M.; Bengtsson, A.A. Degradation of neutrophil extracellular traps co-varies with disease activity in patients with systemic lupus erythematosus. *Arthritis Res. Ther.* **2013**, *15*, R84. [CrossRef] [PubMed]

52. Leffler, J.; Stojanovich, L.; Shoenfeld, Y.; Bogdanovic, G.; Hesselstrand, R.; Blom, A.M. Degradation of neutrophil extracellular traps is decreased in patients with antiphospholipid syndrome. *Clin. Exp. Rheumatol.* **2014**, *32*, 66–70. [PubMed]

53. Kambas, K.; Mitroulis, I.; Apostolidou, E.; Girod, A.; Chrysanthopoulou, A.; Pneumatikos, I.; Skendros, P.; Kourtzelis, I.; Koffa, M.; Kotsianidis, I.; et al. Autophagy mediates the delivery of thrombogenic tissue factor to neutrophil extracellular traps in human sepsis. *PLoS ONE* **2012**, *7*, e45427. [CrossRef] [PubMed]

54. Kambas, K.; Mitroulis, I.; Ritis, K. The emerging role of neutrophils in thrombosis-the journey of TF through NETs. *Front. Immunol.* **2012**, *3*, 385. [CrossRef] [PubMed]

55. Kambas, K.; Chrysanthopoulou, A.; Vassilopoulos, D.; Apostolidou, E.; Skendros, P.; Girod, A.; Arelaki, S.; Froudarakis, M.; Nakopoulou, L.; Giatromanolaki, A.; et al. Tissue factor expression in neutrophil extracellular traps and neutrophil derived microparticles in antineutrophil cytoplasmic antibody associated vasculitis may promote thromboinflammation and the thrombophilic state associated with the disease. *Ann. Rheum. Dis.* **2014**, *73*, 1854–1863. [CrossRef] [PubMed]

56. Brill, A.; Fuchs, T.A.; Savchenko, A.S.; Thomas, G.M.; Martinod, K.; De Meyer, S.F.; Bhandari, A.A.; Wagner, D.D.; et al. Neutrophil extracellular traps promote deep vein thrombosis in mice. *J. Thromb. Haemost.* **2012**, *10*, 136–144. [CrossRef] [PubMed]

57. Stakos, D.A.; Kambas, K.; Konstantinidis, T.; Mitroulis, I.; Apostolidou, E.; Arelaki, S.; Tsironidou, V.; Giatromanolaki, A.; Skendros, P.; Konstantinides, S.; et al. Expression of functional tissue factor by neutrophil extracellular traps in culprit artery of acute myocardial infarction. *Eur. Heart J.* **2015**, *36*, 1405–1414. [CrossRef] [PubMed]

58. von Brühl, M.L.; Stark, K.; Steinhart, A.; Chandraratne, S.; Konrad, I.; Lorenz, M.; Khandoga, A.; Tirniceriu, A.; Coletti, R.; Köllnberger, M.; et al. Monocytes, neutrophils, and platelets cooperate to initiate and propagate venous thrombosis in mice in vivo. *J. Exp. Med.* **2012**, *209*, 819–835. [CrossRef] [PubMed]

59. Pinegin, B.; Vorobjeva, N.; Pinegin, V. Neutrophil extracellular traps and their role in the development of chronic inflammation and autoimmunity. *Autoimmun. Rev.* **2015**, *14*, 633–640. [CrossRef] [PubMed]

60. Engelmann, B.; Massberg, S. Thrombosis as an intravascular effector of innate immunity. *Nat. Rev. Immunol.* **2013**, *13*, 34–45. [CrossRef] [PubMed]

61. Redecha, P.; Tilley, R.; Tencati, M.; Salmon, J.E.; Kirchhofer, D.; Mackman, N.; Girardi, G. Tissue factor: A link between C5a and neutrophil activation in antiphospholipid antibody induced fetal injury. *Blood* **2007**, *110*, 2423–2431. [CrossRef] [PubMed]

62. Fay, W.P. Linking inflammation and thrombosis: Role of C-reactive protein. *World J. Cardiol.* **2010**, *2*, 365–369. [CrossRef] [PubMed]

63. Ritis, K.; Doumas, M.; Mastellos, D.; Micheli, A.; Giaglis, S.; Magotti, P.; Rafail, S.; Kartalis, G.; Sideras, P.; Lambris, J.D.; et al. A Novel C5a Receptor-Tissue Factor Cross-Talk in Neutrophils Links Innate Immunity to Coagulation Pathways. *J. Immunol.* **2006**, *177*, 4794–4802. [CrossRef] [PubMed]

64. Girardi, G.; Mackman, N. Chapter 5 Tissue Factor in Antiphospholipid Antibody-induced Pregnancy Loss: Thrombosis versus Inflammation. *Handb. Syst. Autoimmune Dis.* **2009**, *10*, 69–79. [CrossRef]

65. Galarza-Maldonado, C.; Kourilovitch, M.R.; Andrade-Sánchez, P.; Durán, M.C.; Asanza, E. Treating obstetric antiphospholipid syndrome. *Int. J. Clin. Rheumtol.* **2013**, *8*, 407–414. [CrossRef]

66. Lapponi, M.J.; Carestia, A.; Landoni, V.I.; Rivadeneyra, L.; Etulain, J.; Negrotto, S.; Pozner, R.G.; Schattner, M. Regulation of neutrophil extracellular trap formation by anti-inflammatory drugs. *J. Pharmacol. Exp. Ther.* **2013**, *345*, 430–437. [CrossRef] [PubMed]

67. Fuchs, T.A.; Brill, A.; Duerschmied, D.; Schatzberg, D.; Monestier, M.; Myers, D.D., Jr.; Wrobleski, S.K.; Wakefield, T.W.; Hartwig, J.H.; Wagner, D.D. Extracellular DNA traps promote thrombosis. *Proc. Natl. Acad. Sci. USA* **2010**, *107*, 15880–15885. [CrossRef] [PubMed]

68. Li, L.; Ling, Y.; Huang, M.; Yin, T.; Gou, S.M.; Zhan, N.Y.; Xiong, J.X.; Wu, H.S.; Yang, Z.Y.; Wang, C.Y. Heparin inhibits the inflammatory response induced by LPS and HMGB1 by blocking the binding of HMGB1 to the surface of macrophages. *Cytokine* **2015**, *72*, 36–42. [CrossRef] [PubMed]

69. Manfredi, A.A.; Rovere-Querini, P.; D'Angelo, A.; Maugeri, N. Low molecular weight heparins prevent the induction of autophagy of activated neutrophils and the formation of neutrophil extracellular traps. *Pharmacol. Res.* **2017**. [CrossRef] [PubMed]

70. Barnado, A.; Crofford, L.J.; Oates, J.C. At the Bedside: Neutrophil extracellular traps (NETs) as targets for biomarkers and therapies in autoimmune diseases. *J. Leukoc. Biol.* **2016**, *99*, 265–278. [CrossRef] [PubMed]

71. Müller-Calleja, N.; Manukyan, D.; Canisius, A.; Strand, D.; Lackner, K.J. Hydroxychloroquine inhibits proinflammatory signalling pathways by targeting endosomal NADPH oxidase. *Ann. Rheum. Dis.* **2016**. [CrossRef] [PubMed]

72. Eskelinen, E.L.; Saftig, P. Autophagy: A lysosomal degradation pathway with a central role in health and disease. *Biochim. Biophys. Acta* **2009**, *1793*, 664–673. [CrossRef] [PubMed]

73. Calton, E.K.; Keane, K.N.; Newsholme, P.; Soares, M.J. The impact of Vitamin D levels on inflammatory status: A systematic review of immune cell studies. *PLoS ONE* **2015**, *10*, 1–12. [CrossRef] [PubMed]

74. Handono, K.; Sidarta, Y.O.; Pradana, B.A.; Nugroho, R.A.; Hartono, I.A.; Kalim, H.; Endharti, A.T. Vitamin D prevents endothelial damage induced by increased neutrophil extracellular traps formation in patients with systemic lupus erythematosus. *Acta Med. Indones.* **2014**, *46*, 189–198. [PubMed]

75. Mohammed, B.M.; Fisher, B.J.; Kraskauskas, D.; Farkas, D.; Brophy, D.F.; Fowler, A.A., 3rd; Natarajan, R. Vitamin C: A novel regulator of neutrophil extracellular trap formation. *Nutrients* **2013**, *5*, 3131–3151. [CrossRef] [PubMed]

76. Leshner, M.; Wang, S.; Lewis, C.; Zheng, H.; Chen, X.A.; Santy, L.; Wang, Y. PAD4 mediated histone hypercitrullination induces heterochromatin decondensation and chromatin unfolding to form neutrophil extracellular trap-like structures. *Front. Immunol.* **2012**, *3*, 1–11. [CrossRef] [PubMed]

77. Vivekananthan, D.P.; Penn, M.S.; Sapp, S.K.; Hsu, A.; Topol, E.J. Use of antioxidant vitamins for the prevention of cardiovascular disease: Meta-analysis of randomised trials. *Lancet* **2003**, *361*, 2017–2023. [CrossRef]

78. Gluud, L.L.; Simonetti, R.G. Mortality in Randomized Trials of Antioxidant Supplements for. *J. Am. Med. Assoc.* **2007**, *297*, 844–857. [CrossRef]

79. Sesso, H.D. Vitamins E and C in the Prevention of Cardiovascular Disease in Men. *JAMA* **2008**, *300*, 2123. [CrossRef] [PubMed]

80. Khandpur, R.; Carmona-Rivera, C.; Vivekanandan-Giri, A.; Gizinski, A.; Yalavarthi, S.; Knight, J.S.; Friday, S.; Li, S.; Patel, R.M.; Subramanian, V.; et al. NETs are a source of citrullinated autoantigens and stimulate inflammatory responses in rheumatoid arthritis. *Sci. Transl. Med.* **2013**, *5*. [CrossRef] [PubMed]

81. Alijotas-Reig, J.; Esteve-Valverde, E.; Ferrer-Oliveras, R.; Llurba, E.; Gris, J.M. Tumor Necrosis Factor-Alpha and Pregnancy: Focus on Biologics. An Updated and Comprehensive Review. *Clin. Rev. Allergy Immunol.* **2017**. [CrossRef] [PubMed]

82. Gómez-Puerta, J.A.; Cervera, R. Are there additional options for the treatment of refractory obstetric antiphospholipid syndrome? *Lupus* **2013**, *22*, 754–755. [CrossRef] [PubMed]

83. Mahadevan, U.; Wolf, D.C.; Dubinsky, M.; Cortot, A.; Lee, S.D.; Siegel, C.A.; Ullman, T.; Glover, S.; Valentine, J.F.; Rubin, D.T.; et al. Placental Transfer of Anti–Tumor Necrosis Factor Agents in Pregnant Patients With Inflammatory Bowel Disease. *Clin. Gastroenterol. Hepatol.* **2013**, *11*, 286–292. [CrossRef] [PubMed]

84. Danesh, F.R.; Anel, R.L.; Zeng, L.; Lomasney, J.; Sahai, A.; Kanwar, Y.S. Immunomodulatory effects of HMG-CoA reductase inhibitors. *Arch. Immunol. Ther. Exp.* **2003**, *51*, 139–148.

85. Ferrara, D.E.; Swerlick, R.; Casper, K.; Meroni, P.L.; Vega-Ostertag, M.E.; Harris, E.N.; Pierangeli, S.S. Fluvastatin inhibits up-regulation of tissue factor expression by antiphospholipid antibodies on endothelial cells. *J. Thromb. Haemost.* **2004**, *2*, 1558–1563. [CrossRef] [PubMed]

86. Meroni, P.L.; Raschi, E.; Testoni, C.; Tincani, A.; Balestrieri, G.; Molteni, R.; Khamashta, M.A.; Tremoli, E.; Camera, M. Statins prevent endothelial cell activation induced by antiphospholipid (anti-beta2-glycoprotein I) antibodies: Effect on the proadhesive and proinflammatory phenotype. *Arthritis Rheum.* **2001**, *44*, 2870–2878. [CrossRef]

87. López-Pedrera, C.; Ruiz-Limón, P.; Aguirre, M.Á.; Barbarroja, N.; Pérez-Sánchez, C.; Buendía, P.; Rodriguez-García, I.C.; Rodriguez-Ariza, A.; Collantes-Estevez, E.; Velasco, F.; et al. Global effects of fluvastatin on the prothrombotic status of patients with antiphospholipid syndrome. *Ann. Rheum. Dis.* **2011**, *70*, 675–682. [CrossRef]

88. Chow, O.A.; von Köckritz-Blickwede, M.; Bright, A.T.; Hensler, M.E.; Zinkernagel, A.S.; Cogen, A.L.; Gallo, R.L.; Monestier, M.; Wang, Y.; Glass, C.K.; et al. Statins Enhance Formation of Phagocyte Extracellular Traps. *Cell Host Microbe* **2010**, *8*, 445–454. [CrossRef] [PubMed]

89. Haendeler, J.; Hoffmann, J.; Zeiher, A.M.; Dimmeler, S. Antioxidant effects of statins via S-nitrosylation and activation of thioredoxin in endothelial cells: A novel vasculoprotective function of statins. *Circulation* **2004**, *110*, 856–861. [CrossRef] [PubMed]

90. Greenwood, H.; Patel, J.; Mahida, R.; Wang, Q.; Parekh, D.; Dancer, R.C.; Khiroya, H.; Sapey, E.; Thickett, D.R. Simvastatin to modify neutrophil function in older patients with septic pneumonia (SNOOPI): Study protocol for a randomised placebo-controlled trial. *Trials* **2014**, *15*, 332. [CrossRef] [PubMed]

91. John, S.G.; Thorn, J.; Sobonya, R. Statins as a Potential Risk Factor for Autoimmune Diseases. *Am. J. Ther.* **2014**, *21*, e94–e96. [CrossRef] [PubMed]

92. Mohassel, P.; Mammen, A.L. Statin-associated autoimmune myopathy and anti-HMGCR autoantibodies. *Muscle Nerve* **2013**, *48*, 477–483. [CrossRef] [PubMed]

93. Musset, L.; Allenbach, Y.; Benveniste, O.; Boyer, O.; Bossuyt, X.; Bentow, C.; Phillips, J.; Mammen, A.; van Damme, P.; Westhovens, R.; et al. Anti-HMGCR antibodies as a biomarker for immune-mediated necrotizing myopathies: A history of statins and experience from a large international multi-center study. *Autoimmun. Rev.* **2016**, *15*, 983–993. [CrossRef] [PubMed]

94. Zapantis, E.; Furie, R.; Horowitz, D. THU0400 Response to Eculizumab in the Antiphospholipid Antibody Syndrome. *Ann. Rheum. Dis.* **2015**, *74*. [CrossRef]

95. Matsunaga, N.; Tsuchimori, N.; Matsumoto, T.; Ii, M. TAK-242 (resatorvid), a small-molecule inhibitor of Toll-like receptor (TLR) 4 signaling, binds selectively to TLR4 and interferes with interactions between TLR4 and its adaptor molecules. *Mol. Pharmacol.* **2011**, *79*, 34–41. [CrossRef] [PubMed]

antibodies

MDPI

Article

Dimerized Domain V of Beta2-Glycoprotein I Is Sufficient to Upregulate Procoagulant Activity in PMA-Treated U937 Monocytes and Require Intact Residues in Two Phospholipid-Binding Loops

Alexey Kolyada, David A. Barrios and Natalia Beglova *

Department of Medicine, Beth Israel Deaconess Medical Center and Harvard Medical School, Boston, MA 02215, USA; akolyada@yahoo.com (A.K.); dabarrio@bidmc.harvard.edu (D.A.B.)
* Correspondence: nbeglova@yahoo.com; Tel.: +1-617-735-4009

Academic Editor: Ricard Cervera
Received: 23 December 2016; Accepted: 3 May 2017; Published: 2 June 2017

Abstract: Upregulation of the procoagulant activity of monocytes by antibodies to beta2-glycoprotein I (β2GPI) is one of the mechanisms contributing to thrombosis in antiphospholipid syndrome. Current knowledge about receptors responsible for the upregulation of procoagulant activity by β2GPI/anti-β2GPI complexes and their binding sites on β2GPI is far from complete. We quantified the procoagulant activity expressed by phorbol 12-myristate 13-acetate (PMA)-differentiated U937 cells by measuring clotting kinetics in human plasma exposed to stimulated cells. Cells stimulated with anti-β2GPI were compared to cells treated with dimerized domain V of β2GPI (β2GPI-DV) or point mutants of β2GPI-DV. We demonstrated that dimerized β2GPI-DV is sufficient to induce procoagulant activity in monocytes. Using site-directed mutagenesis, we determined that the phospholipid-binding interface on β2GPI is larger than previously thought and includes Lys308 in β2GPI-DV. Intact residues in two phospholipid-binding loops of β2GPI-DV were important for the potentiation of procoagulant activity. We did not detect a correlation between the ability of β2GPI-DV variants to bind ApoER2 and potentiation of the procoagulant activity of cells. The region on β2GPI inducing procoagulant activity in monocytes can now be narrowed down to β2GPI-DV. The ability of β2GPI-DV dimers to come close to cell membrane and attach to it is important for the stimulation of procoagulant activity.

Keywords: beta2-glycoprotein I; antiphospholipid syndrome; antiphospholipid antibodies; anticardiolipin antibody

1. Introduction

Antiphospholipid syndrome (APS) is an autoimmune disease characterized by clinical thrombosis, recurrent fetal loss during pregnancy and the presence of antiphospholipid antibodies [1,2]. Antiphospholipid antibodies (aPL) detected by laboratory tests for APS are highly heterogeneous even in a single patient [3,4]. The majority of aPL recognize serum proteins that bind anionic phospholipids. Autoantibodies that bind directly to anionic phospholipids are often present in diseases that do not have any link to thrombosis and are generally considered irrelevant to APS [5–7]. Nevertheless, it was recently demonstrated that APS patients may have antibodies that bind cardiolipin without serum protein cofactor, and these antibodies are prothrombotic in mice [8]. The heterogeneity of antiphospholipid antibodies and the wide range of clinical features in APS patients suggest that there are multiple pathways leading to the disease [9–11].

There is a wealth of data demonstrating that anti-β2GPI antibodies are common in APS patients and that these antibodies correlate with thrombosis [12–17]. Anti-β2GPI antibodies potentiate thrombus formation in animal models of thrombosis and induce a prothrombotic state in monocytes, platelets and endothelial cells in vitro [18–20]. B2GPI/anti-β2GPI complexes have been reported to interact with several receptors and cell-surface molecules, such as toll-like receptors TLR2, TLR4, TLR8, ApoER2, GPIbα and annexin A2 [21]. The involvement of TLR4, ApoER2 and annexin A2 in the prothrombotic effects of anti-β2GPI antibodies is supported by in vivo studies using murine thrombosis models [22–25]. The relative contribution of these receptors to cellular activation by anti-β2GPI antibodies and the onset of thrombosis in vivo remains poorly understood. It was recently demonstrated in monocytes that endocytosis is required for anti-β2GPI signaling [26].

B2GPI is a serum protein consisting of five domains [27]. Flexible linkers between domains permit the β2GPI molecule to adopt different shapes. The circular shape in which domain I is adjacent to domain V is the predominant conformation of β2GPI in normal human plasma. In the pathologically-active extended conformation of β2GPI, domain V is independent of other β2GPI domains. Anti-β2GPI antibodies in APS patients with thrombosis most often bind to domain I of β2GPI [28]. Current knowledge of how β2GPI/antibody complexes interact with receptors is incomplete. It is limited to ApoER2, GPIbα and anionic phospholipids. The binding sites for these receptors were localized to domain V of β2GPI [29–34].

Induction of tissue factor (TF) in endothelial cells and monocytes is an important prothrombotic mechanism of β2GPI/anti-β2GPI complexes [35–37]. Monocytes isolated from APS patients have elevated expression of TF and TF-dependent procoagulant activity [38–40]. The ability of patients' IgG to stimulate TF activity in monocytes in vitro correlates with the presence of clinical thrombosis and the levels of anti-β2GPI antibodies in IgG samples [41]. Experimental data implicate TLR2, TLR4, ApoER2 and annexin A2 in the upregulation of TF by anti-β2GPI antibodies [23,42–44]. The binding site on β2GPI for the receptor responsible for the induction of procoagulant activity is unknown. Indefinite anticoagulation, which is a treatment of choice for high risk APS patients, is not always effective in preventing the recurrence of thrombosis [45–47]. A detailed understanding of how β2GPI interacts with receptors involved in cellular activation by β2GPI/anti-β2GPI complexes is essential for the development of drugs specific for antiphospholipid syndrome.

In this study, we compared dimerized β2GPI-DV to β2GPI/anti-β2GPI complexes by its ability to stimulate procoagulant activity in phorbol 12-myristate 13-acetate (PMA)-differentiated U937 cells. U937 is a monocytic cell line, where cells are arrested at an early stage of differentiation. Treatment with PMA induces differentiation of U937 cells to monocytes/macrophages characterized by expression of CD14 and CD11a, CD11b and CD18 integrins [48]. U937 monocytes express all receptors that were suggested to interact with β2GPI/anti-β2GPI complexes [49–52] and respond to antibodies isolated from APS patients with thrombosis by upregulating TF [41]. We used site-directed mutagenesis to change residues in β2GPI-DV involved in binding to ApoER2 and anionic phospholipids and compared how these mutations affected the induction of procoagulant activity expressed by PMA-differentiated U937 cells.

2. Results

2.1. In the Presence of Dimerizing Antibodies, Domain V of β2GPI Is Sufficient to Stimulate the Procoagulant Activity of PMA-Differentiated U937 Monocytes

To mimic dimerized domain V in β2GPI/anti-β2GPI complexes, we attached an HA-tag (amino acid sequence YPYDVPDYA) at the N-terminus of domain V and used anti-HA-tag antibodies to form HA-DV dimers. The procoagulant activity induced by HA-DV/anti-HA complexes in U937 cells was compared to that induced by β2GPI/anti-β2GPI complexes (Figure 1A,B). The procoagulant activity was quantified using coagulation kinetics curves (Figure 1B). Each experimental condition was characterized by the time required to achieve half-maximal coagulation. In our preliminary studies, we performed dose response experiments to determine the concentrations of anti-β2GPI and anti-HA antibodies that are necessary to induce the same level of procoagulant activity in cells as the procoagulant activity induced by 1 μg/mL of LPS. We found the needed concentrations to be approximately 100 nM, which is what we used in our studies. The concentrations of β2GPI and HA-DV (~0.4 μM of β2GPI in 10% serum and 0.8 μM of HA-DV) were in excess of the concentrations of anti-β2GPI and anti-HA antibodies used, so that antibodies were fully loaded with the antigen.

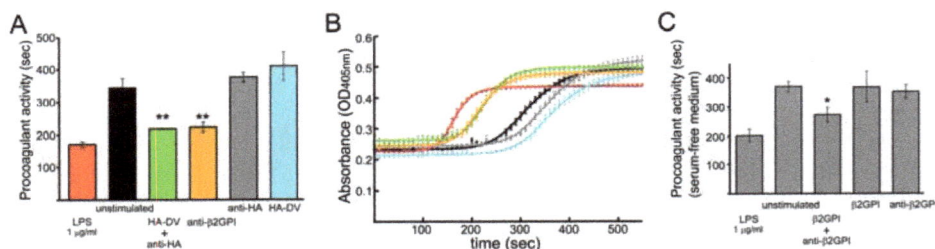

Figure 1. Procoagulant activity induced in U937 cells. (**A**) Cell culture medium contained 10% normal human serum, which was a source of β2GPI. PMA-treated U937 cells were incubated for 6 h with LPS (1 μg/mL); medium; HA-DV (8 μg/mL) with anti-HA (14 μg/mL); anti-β2GPI (16 μg/mL); anti-HA (14 μg/mL) alone; and HA-DV (8 μg/mL) alone. ** $p < 0.001$ compared to medium, HA-DV only and anti-HA only; (**B**) Example of coagulation kinetics curves measured in one of the three experiments used to quantify procoagulant activity in (A). From left to right are the kinetics curves corresponding to cells treated with LPS (red); HA-DV with anti-HA (green); anti-β2GPI (orange); medium (black); anti-HA alone (gray); and HA-DV alone (cyan). Each data point represents the mean and the deviation from the mean of measurements performed in duplicates; (**C**) Procoagulant activity induced by anti-β2GPI in serum-free medium depends on the presence of β2GPI. PMA-treated U937 cells were incubated for 6 h in serum-free medium supplemented with LPS (1 μg/mL); medium; β2GPI (20 μg/mL) with anti-β2GPI (16 μg/mL); β2GPI (20 μg/mL) alone; and anti-β2GPI (16 μg/mL) alone. * $p < 0.05$ compared to medium and β2GPI alone. Procoagulant activities in (A,C) were quantified from coagulation kinetics curves and expressed as time to half-maximal coagulation. Values represent mean ± SD ($n = 3$).

Normal human plasma exposed to cells treated with anti-β2GPI or anti-HA-tag antibodies in the presence of HA-DV coagulated significantly faster than plasma exposed to untreated cells (Figure 1A). Treating cells with either HA-DV alone or anti-HA antibodies alone did not change coagulation time compared to untreated cells. The acceleration of coagulation by monocytes stimulated with anti-β2GPI antibodies was β2GPI dependent (Figure 1C). When cells were stimulated under serum-free conditions, only cells exposed to both β2GPI and anti-β2GPI antibodies significantly accelerated coagulation compared to untreated cells. Neither anti-β2GPI antibodies nor β2GPI alone had any effect on coagulation time in a serum-free medium.

2.2. Measured Procoagulant Activity of U937 Cells Is TF-Dependent

The coagulation cascade consists of two pathways leading to the formation of the fibrin clot: the intrinsic and extrinsic pathways. The intrinsic clotting pathway is activated by the contact activation of FXII, and the extrinsic clotting pathway is initiated by the TF/FVIIa complex. To differentiate the contribution of FXII-dependent and TF-dependent pathways to the initiation of the measured procoagulant activity of the treated U937 cells, we used plasmas deficient in factors FVII, FXII and FXI. Deficient plasmas were exposed to U937 cells stimulated for 6 h with test reagents. The coagulation kinetics of deficient plasmas were compared to the coagulation kinetics of normal plasma (Figure 2). For all stimulants (LPS, HA-DV/anti-HA complexes, anti-β2GPI and untreated cells), factor FVII-deficient plasma exposed to stimulated cells clotted significantly more slowly than normal plasma. The absence of factors FXI or FXII had no effect on the clotting kinetics, when compared to the clotting kinetics of normal plasma. These results demonstrated that the TF/FVIIa complex formed on the surface of U937 cells was the major activator of the clotting cascade in our experiments. Therefore, our assay detects the procoagulant activity of cell-surface TF, which is upregulated by the treatment with anti-β2GPI antibodies and with dimerized β2GPI-DV (Figure 1A).

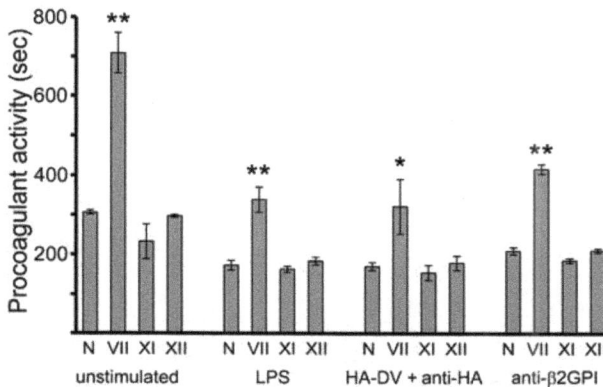

Figure 2. Procoagulant activity exhibited by U937 cells is caused by the upregulation of cell-surface tissue factor. Cell culture medium contained 10% normal human serum. PMA-treated U937 cells were stimulated for 6 h with medium only; LPS (1 µg/mL); HA-DV (8 µg/mL) with anti-HA (14 µg/mL); and anti-β2GPI (16 µg/mL). The procoagulant activities of the cells were measured in pooled normal human plasma (N), as well as in plasmas deficient in factors VII, XI and XII. ** $p < 0.001$, * $p < 0.01$, compared to normal plasma.

2.3. TNFα Released by U937 Cells Stimulated with either HA-DV/Anti-HA Complexes or Anti-β2GPI Antibodies Was Negligible Compared to TNFα Released by Cells Stimulated with TLR4 and TLR2 Ligands

We measured the amount of TNFα released by cells stimulated with either anti-β2GPI antibodies or HA-DV/anti-HA complexes and compared it to the amounts of TNFα secreted by cells stimulated with the TLR4-specific ligand LPS and the TLR2-specific ligand Pam3CSK4 (Figure 3A–D). LPS and Pam3CSK4 upregulated procoagulant activity and induced a massive release of TNFα from U937 cells (Figure 3C,D). Interestingly, although the procoagulant activity induced by anti-β2GPI antibodies and by HA-DV/anti-HA complexes was similar to that induced by LPS, neither anti-β2GPI antibodies nor HA-DV/anti-HA complexes induced appreciable release of TNFα from U937 cells.

2.4. Design and Characterization of Point Mutants of Domain V of β2GPI (HA-DV)

Information, detailed at the amino acid resolution, on how β2GPI interacts with cells is limited to ApoER2 and anionic phospholipids [29,30,32,34,53]. Domain V of β2GPI contains residues critical for the binding to ApoER2 and anionic phospholipids (Figure 4). U937 cells express two isoforms of ApoER2 [52], each of which contains the β2GPI-binding module A1 in the ligand-binding domain [54]. We made point mutants of the HA-tagged domain V of β2GPI (HA-DV) with the goal of dissecting the contribution of ApoER2 and anionic phospholipids to potentiation of procoagulant activity in monocytes treated with dimerized HA-DV. The selected residues were Lys308 and Lys282, which are involved in the binding of β2GPI domain V to ApoER2 [34,53] and the residues in two phospholipid-binding loops (Figure 4). One of the phospholipid-binding loops contains basic residues Lys284, Lys286 and Lys287, and the other loop is composed of a hydrophobic sequence between the residues Leu313 and Trp316 [29,30,32].

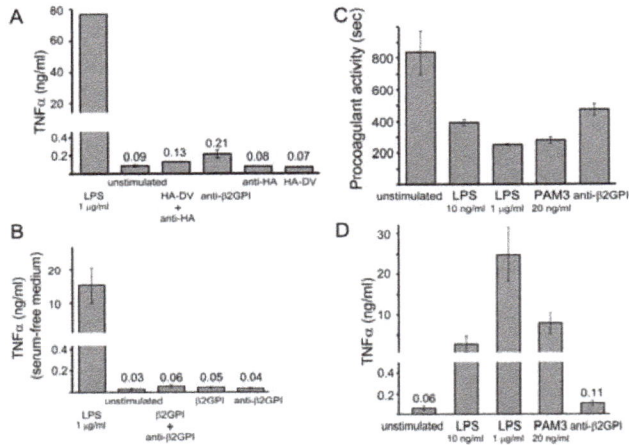

Figure 3. TNFα released into the medium from U937 cells. (**A**) Cell culture medium contained 10% normal human serum. PMA-treated U937 cells were incubated for 6 h with LPS (1 µg/mL); medium; HA-DV (8 µg/mL) with anti-HA (14 µg/mL); anti-β2GPI (16 µg/mL); anti-HA (14 µg/mL) alone; and HA-DV (8 µg/mL) alone; (**B**) PMA-treated U937 cells were incubated for 6 h in a serum-free medium supplemented with LPS (1 µg/mL); medium; β2GPI (20 µg/mL) with anti-β2GPI (16 µg/mL); β2GPI (20 µg/mL) alone; and anti-β2GPI (16 µg/mL) alone. Values represent mean ± SD ($n = 3$); (**C,D**) Procoagulant activities (C) and TNFα released into the medium (D). PMA-treated U937 cells were incubated for 6 h with medium; LPS (10 ng/mL); LPS (1 µg/mL); Pam3CSK4 (20 ng/mL); and anti-β2GPI (16 µg/mL). Results are expressed as the mean ± deviation from the mean ($n = 2$). Mean value of released TNFα is specified at the top of a column.

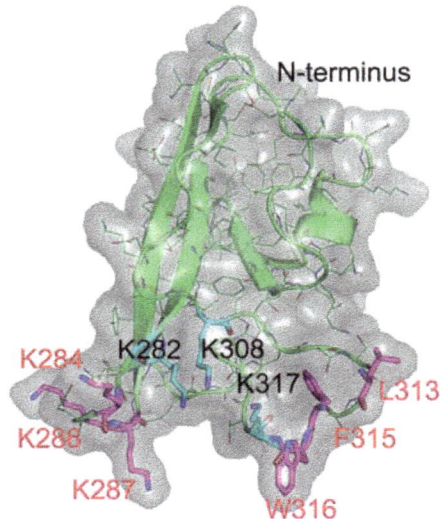

Figure 4. Structure of domain V of β2GPI (cartoon representation). The transparent molecular surface of domain V of β2GPI is colored gray. The residues interacting with A1 (K308, K317 and K282, colored cyan) and the residues in two phospholipid-binding loops (K284, K286, K287 and L313, F315, W316, colored magenta) are rendered as sticks.

2.4.1. The Binding of HA-DV Variants to ApoER2

A1 is a polypeptide that closely resembles the β2GPI-binding module from ApoER2 [53,54]. The ability of HA-DV variants to bind ApoER2 was evaluated by comparing their ability to bind A1. The binding affinity between HA-DV variants and A1 was evaluated by isothermal titration calorimetry (ITC) (Figure 5). ITC is used to directly measure the heat released or absorbed, when binding occurs. HA-DV or HA-DV variants were placed in a sample cell and titrated with A1. Heat changes were detected and measured. First, we measured the binding curve and calculated the binding constant for the HA-DV/A1 complex. We then used the same experimental conditions to compare HA-DV variants to HA-DV with respect to their ability to bind A1. The quantity of heat released upon binding, which is measured by ITC, is directly proportional to the amount of binding. We compared titration curves measured for HA-DV mutants to the titration curve measured for HA-DV. Four mutants of HA-DV bound A1 with affinity similar to that of wild type HA-DV. These were the two HA-DV variants with conservative Lys to Arg mutations (Lys308/Arg and Lys282/Arg) and the two HA-DV variants with mutations in one of the two phospholipid-binding loops (Leu313/Asn and Leu313/Asp_Phe315/Ser). The shallow slope of titration curves measured for all other studied mutations in HA-DV strongly suggests that these mutations disrupted the binding of HA-DV mutants to A1. These results confirmed our previous observations that the hydrophobic loop Leu313-Phe315 is far from the binding interface in the HA-DV/A1 complex [34,53].

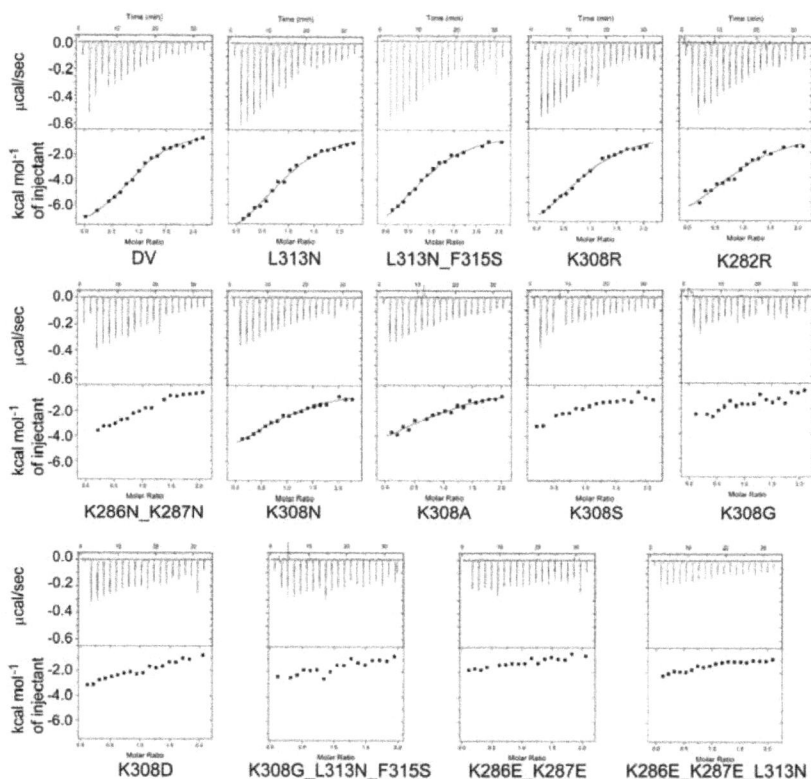

Figure 5. Binding of A1 to HA-DV variants. Isothermal titration calorimetry (ITC) titrations of HA-DV variants (50 μM) in a sample cell with A1 (500 μM) in the injection syringe.

2.4.2. The Binding of HA-DV Variants to Cardiolipin

Next, we analyzed the ability of HA-DV mutants to bind the anionic phospholipid cardiolipin compared to wild type HA-DV. Half-maximal binding to cardiolipin was achieved at a 1.2 μM concentration of wild type HA-DV (Figure 6A). From the cardiolipin-binding curve measured for wild type HA-DV, we selected two concentrations, 500 nM and 1000 nM, which fall in the linear region of the binding curve. The cardiolipin binding ability of HA-DV variants was compared to HA-DV at protein concentrations of 500 nM and 1000 nM (Figure 6B). Only the Lys308/Arg and Lys282/Arg variants of HA-DV retained cardiolipin binding activity similar to that of wild type HA-DV (Figure 6B). Any other residue besides Arg in place of Lys308 dramatically reduced the cardiolipin binding of mutated HA-DV, strongly suggesting that Lys308 is part of the phospholipid-binding interface on β2GPI. Mutations in either of the two phospholipid-binding loops disrupted the binding of HA-DV mutants to cardiolipin, as expected. Three mutants (Lys286/Glu_Lys287/Glu, Lys286/Glu_Lys287/Glu_Leu313/Asn and Lys308/Gly_Leu313/Asn_Phe315/Ser) retained less than 4% of cardiolipin-binding activity compared to wild type HA-DV.

2.5. ApoER2 Does Not Contribute to Upregulation of Procoagulant Activity in U937 Cells

We evaluated how point mutations in domain V affected the ability of HA-DV dimers to stimulate procoagulant activity in U937 cells. As illustrated by Figure 6C, HA-DV variants can be divided into three groups based on their ability to induce procoagulant activity in cells stimulated in the presence of dimerizing anti-HA antibodies. These are (Group 1) HA-DV variants that stimulated cells like wild type HA-DV (the difference in procoagulant activity between unstimulated cells and cells stimulated with HA-DV variants was statistically significant), (Group 2) HA-DV variants that induced procoagulant activity similar to that exhibited by unstimulated cells (the difference in procoagulant activity induced in cells stimulated with wild type HA-DV and cells stimulated with HA-DV variants was statistically significant) and (Group 3) intermediate HA-DV variants, whose activity in cells was not statistically different from either untreated cells or cells stimulated with wild type HA-DV. In the absence of dimerizing anti-HA antibodies, neither of the HA-DV mutants induced procoagulant activity statistically different from procoagulant activity exhibited by unstimulated cells, as we have already demonstrated previously.

HA-DV variants that retained their ability to bind A1 and, therefore, were capable of interacting with ApoER2 (hatched columns on Figure 6C) were distributed among all three groups of HA-DV variants. This result suggests that the binding of HA-DV/anti-HA complexes to ApoER2 is not important for the induction of procoagulant activity in U937 cells.

2.6. Intact Residues in the Two Phospholipid-Binding Loops of HA-DV Are Important for the Ability of HA-DV/Anti-HA Complexes to Induce Procoagulant Activity in U937 Cells

The pathological function of β2GPI is a result of both dimerization of two β2GPI molecules by antibodies and functional interactions with receptors and phospholipids. In our system, the binding of antibodies to the epitope tag creates HA-DV/anti-HA complexes in solution, allowing us to focus on the functional interactions.

It is clear from Figure 6C that mutation in either of the two phospholipid-binding loops in HA-DV resulted in a dimer that does not upregulate procoagulant activity of U937 cells. The ability of mutants to come close to the cell membrane and bind to it, at least to some extent, is important for stimulating the procoagulant activity. Three out of the five mutants that failed to stimulate procoagulant activity (Group 2, Figure 6C) have charge reversal mutations. All five mutants retained less than 20% of the cardiolipin-binding ability of wild type HA-DV, and three of these mutants (gray columns on Figure 6C) retained less than 4% of the cardiolipin binding.

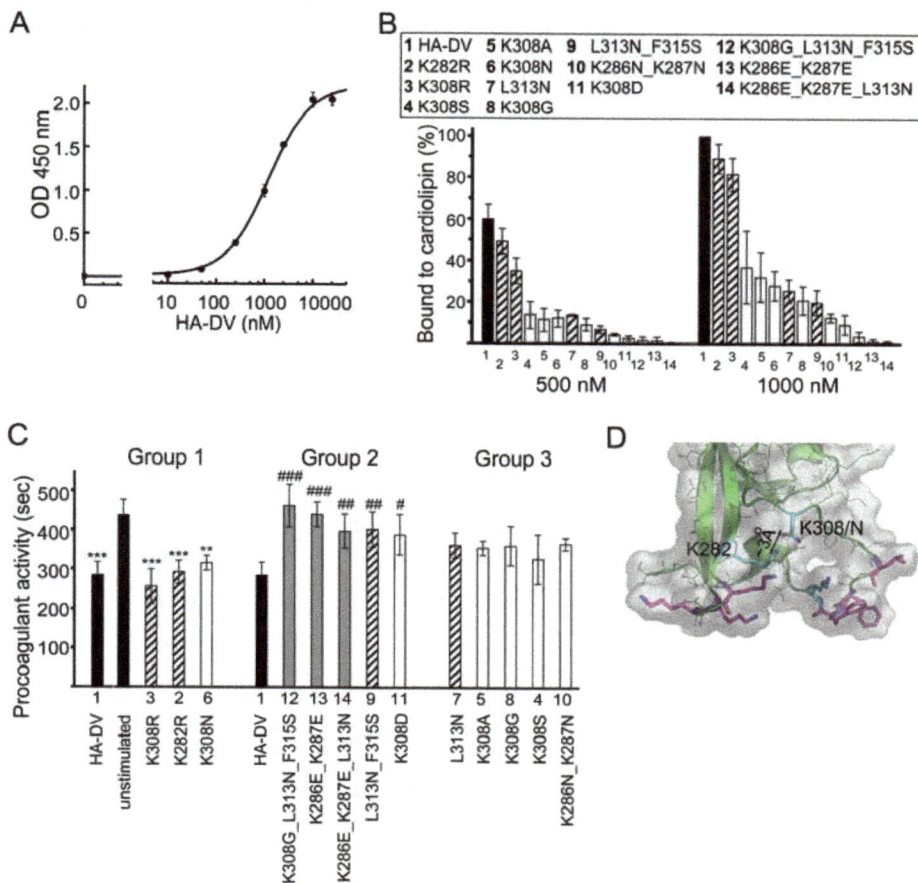

Figure 6. The ability of HA-DV variants to bind cardiolipin and induce procoagulant activity in U937 cells. (**A**) The binding of wild type HA-DV to cardiolipin. The half-maximal binding was achieved at 1.2 μM of HA-DV. Each data point shows the mean ± deviation from the mean of two measurements; (**B**) The ability of HA-DV variants to bind cardiolipin compared to wild type HA-DV (black columns). HA-DV variants that bind A1 are designated by hatched columns. Results are expressed as the percentage of cardiolipin binding measured for wild type HA-DV at 1000 nM. Values represent the mean ± SD ($n = 3$). HA-DV mutants are numbered as in (**C**); (**C**) Procoagulant activity in cells treated with HA-DV variants (8 μg/mL) in the presence of anti-HA (14 μg/mL). *** $p < 0.001$ and ** $p < 0.01$ compared to medium. ### $p < 0.001$, ## $p < 0.01$ and # $p < 0.05$ compared to wild type HA-DV in the presence of anti-HA. HA-DV variants are numbered as in (**B**). HA-DV variants that bind A1 (hatched columns). HA-DV variants that have less than 4% of cardiolipin binding (gray columns). Values represent the mean ± SD ($n = 3$); (**D**) Putative hydrogen bond formed between the sidechains of residues K282 and N308 in the K308/N mutant of HA-DV. The transparent molecular surface of domain V of β2GPI is colored gray. Sidechains interacting with A1 (cyan) and with phospholipids (magenta) are rendered as sticks.

Three of the studied mutants, Lys308/Arg, Lys282/Arg and Lys308/Asn, were as good as wild type HA-DV in stimulating the procoagulant activity in U937 cells (Group 1, Figure 6C). A conservative Lys to Arg mutation often has little effect on protein function explaining why Lys308/Arg and

Lys282/Arg mutants closely resemble wild type HA-DV in stimulating procoagulant activity. On the other hand, the Lys308/Asn mutant showed significantly reduced ability to bind cardiolipin, but retained its ability to stimulate procoagulant activity in monocytes. We found an explanation for this result by analyzing the structure of domain V of β2GPI available in the PDB data bank (PDB ID 1C1Z, 1QUB, 3OP8 and 2KRI). The Lys308/Asn mutant, compared to other less potent Lys308 mutants (Lys308/Ala, Lys308/Gly and Lys308/Ser), is capable of forming a hydrogen bond with the sidechain of Lys282. In the Lys308/Asn mutant, this hydrogen bond combined with phospholipid-bound Leu313 restricts flexibility in the unstructured region between the residues 308 and 313 (Figure 6D). This unstructured region is stabilized by the binding of Lys308 and Leu313 to anionic phospholipids in wild type HA-DV. It is likely that the region between residues 308 and 313 is in the vicinity of the binding site for the receptor, because its flexibility affects the ability of HA-DV dimers to stimulate procoagulant activity in treated cells.

3. Discussion

We demonstrated that domain V of β2GPI (β2GPI-DV) dimerized to mimic domain V in β2GPI/anti-β2GPI complexes is sufficient to induce procoagulant activity in PMA-differentiated U937 monocytic cells. Our data considerably simplify the search for the residues on β2GPI, which are involved in the upregulation of procoagulant activity by anti-β2GPI antibodies. The use of β2GPI-DV dimers instead of a full-length β2GPI can also simplify the search for receptors involved in the upregulation of procoagulant activity in monocytes by anti-β2GPI antibodies. This is a step towards understanding how β2GPI/anti-β2GPI complexes interact with receptors and, ultimately, towards a drug to treat anti-β2GPI-related thrombosis in APS.

Using site-directed mutagenesis, we changed individual residues in β2GPI-DV involved in the binding of β2GPI to ApoER2 and anionic phospholipids and compared the procoagulant activity induced by dimerized β2GPI-DV variants in treated U937 cells. It has been previously shown that domain V of β2GPI is important for stimulating platelet adhesion to collagen by dimeric β2GPI and that the increase of platelet adhesion is mediated by ApoER2 [33,55]. We did not find a correlation between the ability of a β2GPI-DV variant to stimulate procoagulant activity of monocytic cells and its ability to bind A1, which is the β2GPI-binding module from ApoER2. Our results suggest that binding to ApoER2 is not important for stimulating the procoagulant activity of monocytic cells by anti-β2GPI antibodies, highlighting the complexity of molecular mechanisms of thrombosis in antiphospholipid syndrome.

The surface area on β2GPI-DV involved in phospholipid binding is much larger than previously thought. Our data suggest that Lys308 actively participates in binding of β2GPI-DV to anionic phospholipids. When Lys308 was mutated to either Ser, Ala, Asn, Gly or Asp, cardiolipin binding was reduced to 37%, 32%, 28%, 21% and 9% of the level of wild type β2GPI-DV, respectively.

Our results show that the ability of mutants to come close to the cell membrane and attach to it, even moderately, is important for the stimulation of procoagulant activity. All active mutants in Group 1 in Figure 6C retained native residues in both phospholipid-binding loops, compared to inactive mutants in Group 2.

It has been previously shown that the binding of β2GPI to anionic phospholipids has two effects: (1) it causes conformational rearrangement of a full-length β2GPI to expose an epitope for anti-β2GPI antibodies otherwise hidden on β2GPI [56] and (2) creates local density of β2GPI to facilitate the formation of multivalent complexes with low affinity APS antibodies [57–59]. Using antibodies to an unobstructed HA-tag attached to β2GPI-DV, we have shown that the ability of β2GPI-DV to reach the cell membrane and attach to it is important for signaling by β2GPI/anti-β2GPI complexes.

Our data suggest that the binding of β2GPI-DV to anionic phospholipids restricts the flexibility of the unstructured region in β2GPI-DV between the residues 308 and 313 influencing the ability of dimerized domain V of β2GPI to stimulate procoagulant activity in U937 cells. Stabilization of this region is achieved in wild type β2GPI-DV by anchoring Lys308 and Leu313 to a phospholipid membrane. In the Lys308/Asn mutant, which stimulates the procoagulant activity of cells similar to wild type β2GPI-DV, the unstructured region between the residues 308 and 313 is stabilized by a hydrogen bond constraining the Asn308 residue and by phospholipid-bound Leu313. Since the flexibility of the stretch of residues between Lys308 and Leu313 influences the ability β2GPI-DV dimers to stimulate procoagulant activity, this region in β2GPI-DV is likely close to the binding site for a cell-surface receptor.

Our results suggest a model in which β2GPI binds by its domain V to anionic phospholipids on cellular surfaces, most likely to lipid rafts enriched in anionic phospholipids and signaling proteins [60]. The binding to anionic phospholipids restricts flexibility in the unstructured loop between phospholipid-bound residues Lys308 and Leu313 in β2GPI-DV, predisposing β2GPI-DV for binding to a receptor. Anti-β2GPI antibodies keep β2GPI attached to cellular membranes by increasing the avidity of β2GPI/antibody complexes for anionic phospholipids. Binding to a receptor occurs very close to the cellular surface, because β2GPI-DV has to be attached to anionic phospholipids in order to interact with a receptor (Figure 7). Whether the interaction of β2GPI-DV with cell-surface receptors leads directly to the stimulation of procoagulant activity in cells or facilitates the endocytosis of dimerized β2GPI-DV, which then signals from endosomes, awaits further investigation.

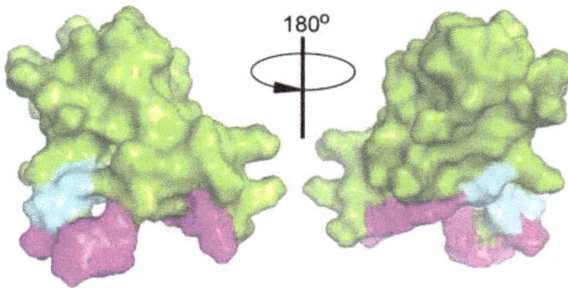

Figure 7. Surface representation of β2GPI-DV. The surface area around residues involved in binding to anionic phospholipids is colored magenta; the unstructured region between Lys308 and Leu313 is colored cyan.

The procoagulant activity induced in monocytes by anti-β2GPI antibodies depends on cell-surface TF. In isolated normal peripheral blood mononuclear cells (PBMC), anti-β2GPI antibodies significantly increased cell-surface TF activity and TF mRNA levels [42,61,62]. We have shown that cell-surface TF is a major contributor to the increased procoagulant activity of PMA-differentiated U937 monocytic cells treated with anti-β2GPI antibodies and β2GPI-DV dimers. The mechanism by which treatment with anti-β2GPI antibodies and β2GPI-DV dimers affects TF in U937 cells is not yet clear. We will continue investigating the extent to which treatment with anti-β2GPI antibodies contributes to de novo synthesis of TF versus decryption of TF already present on the cellular surface. Activation of cell-surface TF by anti-β2GPI antibodies and β2GPI-DV dimers could be accompanied by an increase in surface exposure of anionic phospholipids additionally contributing to procoagulant activity in the treated cells.

It is not yet clear what receptor in monocytes is responsible for the induction of procoagulant activity by anti-β2GPI antibodies and β2GPI-DV dimers. Experiments in PBMC implicate TLR2, TLR4 and TLR8 in the upregulation of TF by anti-β2GPI antibodies, which is accompanied by a TNFα release ranging from 0.4–10 ng/mL [26,42,44,63]. TLR8 is a likely endosomal receptor for β2GPI/anti-β2GPI complexes [63]. It is also possible that endosomes have a not yet identified receptor contributing to monocyte activation by anti-β2GPI antibodies. We found that β2GPI/anti-β2GPI complexes, dimerized β2GPI-DV, LPS and Pam3CSK4 all induced procoagulant activity in PMA-differentiated U937 cells. However, β2GPI/anti-β2GPI complexes and dimerized β2GPI-DV did not promote the release of TNFα, in contrast to LPS and Pam3CSK4, which caused a massive release of TNFα into the cell culture medium. Our results suggest that another receptor, besides TLR4 and TLR2, can contribute to the upregulation of procoagulant activity in monocytes by β2GPI/antibody complexes and that the stimulation of this receptor does not lead to NF-κB activation. More investigation is required into the details of the signaling pathways induced by β2GPI/anti-β2GPI complexes and by dimers of β2GPI-DV in PMA-differentiated U937 cells and how they compare to the signaling pathways induced in PBMC.

In conclusion, our studies in PMA-differentiated U937 monocytes have narrowed the location of the region on β2GPI responsible for the induction of procoagulant activity in monocytes by β2GPI/anti-β2GPI complexes down to domain V. Intact residues in β2GPI-DV that bind to anionic phospholipids are important for the potentiation of procoagulant activity in monocytes. The binding site for a cell-surface receptor on β2GPI-DV is likely located in the vicinity of an unstructured region in β2GPI-DV between residues 308 and 313. The flexibility of this region, which is restricted in phospholipid-bound β2GPI-DV, affects the ability of dimerized β2GPI-DV to stimulate procoagulant activity on monocytes. Our data suggest that ApoER2 is not important for the potentiation of procoagulant activity in PMA-differentiated U937 cells. The identity of the receptor that plays a role in stimulating procoagulant activity in U937 cells and the signaling pathways initiated by β2GPI/anti-β2GPI complexes and β2GPI-DV dimers awaits further investigation.

4. Materials and Methods

4.1. Proteins

A1 is a fragment of mouse ApoER2 (residues 12–47) in which Asp is substituted for Asn36. A1 was expressed in *Escherichia coli* and purified as previously described [64]. HA-DV consists of an HA tag (amino acid sequence YPYDVPDYA) added to the N-terminus of domain V of human β2GPI (residues 244–326). HA-DV was subcloned into a pET15b vector (Novagen) in which the sequence recognized by the tobacco etch virus (TEV) protease was added after an N-terminal histidine tag so that the tag can be removed. The HA-DV protein and point mutants of HA-DV were expressed and purified as previously described [34].

4.2. Cells and Culture Conditions

The immortalized human monocyte U937 cells (ATCC, Manassas, VA, USA) were cultured in RPMI 1640 medium supplemented with 10% fetal bovine serum (FBS) (Atlanta Biologicals, Flowery Branch, GA, USA), penicillin-streptomycin and L-glutamine (Gibco, ThermoFisher Scientific, Waltham, MA, USA) at 37 °C in a humidified atmosphere with 5% CO_2. Cells were seeded at a concentration of 5×10^5 mL^{-1} and treated for 72 h with 100 nM phorbol 12-myristate 13-acetate (PMA) (Enzo Life Sciences, Farmingdale, NY, USA). After 72 h, nonadherent cells were removed along with the medium. A fresh medium containing 10% FBS was added to the cells, and adherent cells were detached by gentle pipetting. Cells were pelleted and resuspended in RPMI medium. Differentiated U937 monocytes at a concentration of 1×10^6 mL^{-1} were incubated for 6 h at 37 °C in a humidified atmosphere with 5% CO_2 in RPMI medium supplemented with 10% pooled normal human serum (Innovative Research, Novi, MI, USA) and test reagents as indicated. Human serum in cell culture media supplied β2GPI. When specified, cells were incubated in a serum free medium with or without purified β2GPI (Haematologic Technologies, Essex Junction, VT, USA), exchanged into a 20 mM Hepes, 150 mM NaCl, pH 7.5 buffer using a Zeba spin desalting column (ThermoFisher Scientific, Waltham, MA, USA) and added to the assay at a final concentration of 20 μg/mL.

HA-DV and HA-DV mutants were used at an 8 μg/mL concentration measured by NanoDrop (ThermoScientific, Wilmington, DE, USA). The TLR2-specific ligand Pam3CSK4 was from InvivoGen, San Diego, CA, USA. Anti-HA antibody (Bethyl Laboratories, Montgomery, TX, USA) was exchanged into a 20 mM Hepes, 150 mM NaCl, pH 7.5 buffer using a Zeba spin desalting column to remove sodium azide. Goat anti-β2GPI (CL2001AP, Cedarlane Laboratories, Burlington, NC, USA) was raised against human β2GPI and affinity purified on immobilized β2GPI. LPS from *Salmonella enterica* (Sigma, St. Louis, MO, USA) was used as a positive control. Endotoxin levels in test reagents were measured with the Limulus Amebocyte Lysate (LAL) chromogenic endotoxin quantification kit (ThermoFisher Scientific, Waltham, MA, USA) at concentrations used in the assays. Endotoxin levels in HA-DV and all HA-DV mutants were below the detection limit of 0.1 $EUmL^{-1}$, except for Lys286/Glu-Lys287/Glu and Lys286/Asn-Lys287/Asn, for which measured endotoxin was 0.5 $EUmL^{-1}$. Endotoxin levels in β2GPI, anti-β2GPI and anti-HA were 0.15, 0.25 and 0.6 $EUmL^{-1}$, respectively. Endotoxin in test reagents was far below 1.5 $EUmL^{-1}$, which corresponds to 1 ng/mL of LPS from *Salmonella enterica*. This amount of LPS did not have a statistically significant effect on U937 cells.

4.3. Measurements of the Procoagulant Activity of U937 Cells

After incubating for 6 h with test reagents, cells were pelleted, washed with RPMI and counted, and their viability, which was at least 90% in reported experiments, was assessed by Trypan Blue (ThermoFisher Scientific, Waltham, MA, USA). The procoagulant activity expressed by cells was quantified by measuring clotting kinetics in pooled normal platelet-poor human plasma anticoagulated with sodium citrate (Innovative Research, Novi, MI, USA). When specified, plasma depleted of factors VII, XI or XII (Haematologic Technologies, Essex Junction, VT, USA) was used in clotting studies. Clotting kinetics were measured at 37 °C using 96-well ELISA plates and a Spectramax 340PC Microplate Reader (Molecular Devices Inc., Sunnyvale, CA, USA). Human plasma, 50 μL, was added to 50 μL of cells (2×10^6 mL^{-1}) suspended in serum-free RPMI. The mixture was incubated for 3 min at 37 °C, and coagulation was initiated by adding 50 μL of 40 mM $CaCl_2$ in 20 mM Hepes, 150 mM NaCl buffer, pH 7.5. Clotting kinetics were recorded by measuring absorbance at 405 nm. Kinetics data were fitted to a 4-parameter equation using the Gnuplot 5.0 program (http://www.gnuplot.info/). The time needed to achieve a half-maximal increase in OD was calculated for each kinetics curve and used to characterize the procoagulant activity of the cells.

4.4. TNFα ELISA

After 6 h of stimulation with test reagents, cells were pelleted, and the supernatant was collected and stored frozen at −80 °C until use. The concentration of TNFα released into media was quantified with Quantikine ELISA (R&D Systems, Minneapolis, MN, USA).

4.5. Isothermal Titration Calorimetry

To measure the binding between A1 and the HA-tagged domain V of β2GPI (HA-DV) and its variants, lyophilized proteins were resuspended in a 25 mM Hepes, pH 7.1 buffer containing 50 mM NaCl and 2 mM $CaCl_2$ and dialyzed overnight at 4 °C in the same buffer. Measurements were performed at 298 K using a MicroCal iTC200 system (Malvern, Malvern Instruments, U.K.). A1 at a concentration of 500 μM was placed into an injection syringe and titrated in 2 μL increments into a sample cell containing 50 μM of HA-DV or HA-DV variants. Binding isotherms were fit to a one site binding model using the Origin software for ITC.

4.6. Cardiolipin ELISA

ELISA 96-well plates (Costar, Corning, NY, USA) were coated with 50 μL per well of cardiolipin (Sigma, St. Louis, MO, USA) prepared at 200 μg/mL in ethanol and blocked for 2 h with 4% BSA in a 20 mM Tris, 100 mM NaCl buffer at pH 7.4. To generate a binding curve, increasing concentrations of HA-DV were applied to wells. The binding data were fit to a one-site model using the Gnuplot 5.0 program (http://www.gnuplot.info/). Cardiolipin binding by HA-DV variants was compared to cardiolipin binding by HA-DV at protein concentrations of 500 nM and 1000 nM. Bound HA-tagged proteins were detected with HRP-conjugated anti-HA-tag antibody (ab1265, Abcam, Cambridge, MA, USA) using a TMB (3,3',5,5'-tetramethylbenzidine) substrate. Absorbances at 450 nm were measured on a Spectramax 340PC Microplate Reader (Molecular Devices Inc., Sunnyvale, CA, USA).

4.7. Statistical Analysis

Results are presented as the mean ± standard deviation calculated from at least three independent experiments. Statistical significance was evaluated with STATA statistical software (College Station, TX, USA) using one-way ANOVA with Bonferroni correction for multiple comparisons.

Acknowledgments: This work was supported by the R01 HL096693 grant from the National Institute of Health to N. Beglova. Salary support for D. A. Barrios was from NIH Fellowship Training Program T32 HL007917.

Author Contributions: A.K. and N.B. designed the experiments. A.K. and D.A.B. performed the experiments. A.K., D.A.B. and N.B. analyzed the data. N.B. devised the studies and wrote the paper.

Conflicts of Interest: The authors declare no conflict of interest.

References

1. Bertolaccini, M.L.; Amengual, O.; Andreoli, L.; Atsumi, T.; Chighizola, C.B.; Forastiero, R.; de Groot, P.; Lakos, G.; Lambert, M.; Meroni, P.; et al. 14th international congress on antiphospholipid antibodies task force. Report on antiphospholipid syndrome laboratory diagnostics and trends. *Autoimmun. Rev.* **2014**, *13*, 917–930. [CrossRef] [PubMed]

2. Miyakis, S.; Lockshin, M.D.; Atsumi, T.; Branch, D.W.; Brey, R.L.; Cervera, R.; Derksen, R.H.; de Groot, P.G.; Koike, T.; Meroni, P.L.; et al. International consensus statement on an update of the classification criteria for definite antiphospholipid syndrome (APS). *J. Thromb. Haemost.* **2006**, *4*, 295–306. [CrossRef] [PubMed]

3. Giles, I.P.; Isenberg, D.A.; Latchman, D.S.; Rahman, A. How do antiphospholipid antibodies bind beta2-glycoprotein I? *Arthritis Rheum.* **2003**, *48*, 2111–2121. [CrossRef] [PubMed]

4. Lieby, P.; Soley, A.; Levallois, H.; Hugel, B.; Freyssinet, J.M.; Cerutti, M.; Pasquali, J.L.; Martin, T. The clonal analysis of anticardiolipin antibodies in a single patient with primary antiphospholipid syndrome reveals an extreme antibody heterogeneity. *Blood* **2001**, *97*, 3820–3828. [CrossRef] [PubMed]

5. Sorice, M.; Pittoni, V.; Griggi, T.; Losardo, A.; Leri, O.; Magno, M.S.; Misasi, R.; Valesini, G. Specificity of anti-phospholipid antibodies in infectious mononucleosis: A role for anti-cofactor protein antibodies. *Clin. Exp. Immunol.* **2000**, *120*, 301–306. [CrossRef] [PubMed]

6. Wolf, P.; Gretler, J.; Aglas, F.; Auer-Grumbach, P.; Rainer, F. Anticardiolipin antibodies in rheumatoid arthritis: Their relation to rheumatoid nodules and cutaneous vascular manifestations. *Br. J. Dermatol.* **1994**, *131*, 48–51. [CrossRef] [PubMed]

7. Forastiero, R.R.; Martinuzzo, M.E.; Kordich, L.C.; Carreras, L.O. Reactivity to beta 2 glycoprotein I clearly differentiates anticardiolipin antibodies from antiphospholipid syndrome and syphilis. *Thromb. Haemost.* **1996**, *75*, 717–720. [PubMed]

8. Manukyan, D.; Muller-Calleja, N.; Jackel, S.; Luchmann, K.; Monnikes, R.; Kiouptsi, K.; Reinhardt, C.; Jurk, K.; Walter, U.; Lackner, K.J. Cofactor-independent human antiphospholipid antibodies induce venous thrombosis in mice. *J. Thromb. Haemost.* **2016**, *14*, 1011–1020. [CrossRef] [PubMed]

9. Merashli, M.; Noureldine, M.H.; Uthman, I.; Khamashta, M. Antiphospholipid syndrome: An update. *Eur. J. Clin. Investig.* **2015**, *45*, 653–662. [CrossRef] [PubMed]

10. Giannakopoulos, B.; Krilis, S.A. The pathogenesis of the antiphospholipid syndrome. *N. Engl. J. Med.* **2013**, *368*, 1033–1044. [CrossRef] [PubMed]

11. Meroni, P.L.; Chighizola, C.B.; Rovelli, F.; Gerosa, M. Antiphospholipid syndrome in 2014: More clinical manifestations, novel pathogenic players and emerging biomarkers. *Arthritis Res. Ther.* **2014**, *16*, 209. [CrossRef] [PubMed]

12. Lopez, L.R.; Dier, K.J.; Lopez, D.; Merrill, J.T.; Fink, C.A. Anti-beta 2-glycoprotein I and antiphosphatidylserine antibodies are predictors of arterial thrombosis in patients with antiphospholipid syndrome. *Am. J. Clin. Pathol.* **2004**, *121*, 142–149. [CrossRef] [PubMed]

13. Day, H.M.; Thiagarajan, P.; Ahn, C.; Reveille, J.D.; Tinker, K.F.; Arnett, F.C. Autoantibodies to beta2-glycoprotein I in systemic lupus erythematosus and primary antiphospholipid antibody syndrome: Clinical correlations in comparison with other antiphospholipid antibody tests. *J. Rheumatol.* **1998**, *25*, 667–674. [PubMed]

14. Cabiedes, J.; Cabral, A.R.; Alarcon-Segovia, D. Clinical manifestations of the antiphospholipid syndrome in patients with systemic lupus erythematosus associate more strongly with anti-beta 2-glycoprotein-I than with antiphospholipid antibodies. *J. Rheumatol.* **1995**, *22*, 1899–1906. [PubMed]

15. De Laat, B.; Pengo, V.; Pabinger, I.; Musial, J.; Voskuyl, A.E.; Bultink, I.E.; Ruffatti, A.; Rozman, B.; Kveder, T.; de Moerloose, P.; et al. The association between circulating antibodies against domain i of beta2-glycoprotein I and thrombosis: An international multicenter study. *J. Thromb. Haemost.* **2009**, *7*, 1767–1773. [CrossRef] [PubMed]

16. Zhang, S.; Wu, Z.; Chen, S.; Li, J.; Wen, X.; Li, L.; Zhang, W.; Zhao, J.; Zhang, F.; Li, Y. Evaluation of the diagnostic potential of antibodies to beta2-glycoprotein I domain I in chinese patients with antiphospholipid syndrome. *Sci. Rep.* **2016**, *6*, 23839. [CrossRef] [PubMed]

17. De Laat, H.B.; Derksen, R.H.; Urbanus, R.T.; Roest, M.; de Groot, P.G. Beta2-glycoprotein I-dependent lupus anticoagulant highly correlates with thrombosis in the antiphospholipid syndrome. *Blood* **2004**, *104*, 3598–3602. [CrossRef] [PubMed]

18. Willis, R.; Pierangeli, S.S. Anti-beta2-glycoprotein I antibodies. *Ann. N. Y. Acad. Sci.* **2013**, *1285*, 44–58. [CrossRef] [PubMed]

19. Arad, A.; Proulle, V.; Furie, R.A.; Furie, B.C.; Furie, B. Beta(2)-glycoprotein-I autoantibodies from patients with antiphospholipid syndrome are sufficient to potentiate arterial thrombus formation in a mouse model. *Blood* **2011**, *117*, 3453–3459. [CrossRef] [PubMed]

20. Jankowski, M.; Vreys, I.; Wittevrongel, C.; Boon, D.; Vermylen, J.; Hoylaerts, M.F.; Arnout, J. Thrombogenicity of beta 2-glycoprotein I-dependent antiphospholipid antibodies in a photochemically induced thrombosis model in the hamster. *Blood* **2003**, *101*, 157–162. [CrossRef] [PubMed]

21. De Groot, P.G.; Urbanus, R.T. The significance of auto-antibodies against beta2-glycoprotein I. *Blood* **2012**, *120*, 266–274. [CrossRef] [PubMed]

22. Pierangeli, S.S.; Vega-Ostertag, M.E.; Raschi, E.; Liu, X.; Romay-Penabad, Z.; De Micheli, V.; Galli, M.; Moia, M.; Tincani, A.; Borghi, M.O.; et al. Toll-like receptor and antiphospholipid mediated thrombosis: In vivo studies. *Ann. Rheum. Dis.* **2007**, *66*, 1327–1333. [CrossRef] [PubMed]

23. Romay-Penabad, Z.; Aguilar-Valenzuela, R.; Urbanus, R.T.; Derksen, R.H.; Pennings, M.T.; Papalardo, E.; Shilagard, T.; Vargas, G.; Hwang, Y.; de Groot, P.G.; et al. Apolipoprotein E receptor 2 is involved in the thrombotic complications in a murine model of the antiphospholipid syndrome. *Blood* **2011**, *117*, 1408–1414. [CrossRef] [PubMed]

24. Romay-Penabad, Z.; Montiel-Manzano, M.G.; Shilagard, T.; Papalardo, E.; Vargas, G.; Deora, A.B.; Wang, M.; Jacovina, A.T.; Garcia-Latorre, E.; Reyes-Maldonado, E.; et al. Annexin A2 is involved in antiphospholipid antibody-mediated pathogenic effects in vitro and in vivo. *Blood* **2009**, *114*, 3074–3083. [CrossRef] [PubMed]

25. Laplante, P.; Fuentes, R.; Salem, D.; Subang, R.; Gillis, M.A.; Hachem, A.; Farhat, N.; Qureshi, S.T.; Fletcher, C.A.; Roubey, R.A.; et al. Antiphospholipid antibody-mediated effects in an arterial model of thrombosis are dependent on toll-like receptor 4. *Lupus* **2016**, *25*, 162–176. [CrossRef] [PubMed]

26. Brandt, K.J.; Fickentscher, C.; Boehlen, F.; Kruithof, E.K.; de Moerloose, P. NF-kappab is activated from endosomal compartments in antiphospholipid antibodies-treated human monocytes. *J. Thromb. Haemost.* **2014**, *12*, 779–791. [CrossRef] [PubMed]

27. De Groot, P.G.; Meijers, J.C. Beta(2)-glycoprotein I: Evolution, structure and function. *J. Thromb. Haemost.* **2011**, *9*, 1275–1284. [CrossRef] [PubMed]

28. Meroni, P.L. Anti-beta-2 glycoprotein I epitope specificity: From experimental models to diagnostic tools. *Lupus* **2016**, *25*, 905–910. [CrossRef] [PubMed]

29. Hunt, J.; Krilis, S. The fifth domain of beta 2-glycoprotein I contains a phospholipid binding site (cys281-cys288) and a region recognized by anticardiolipin antibodies. *J. Immunol.* **1994**, *152*, 653–659. [PubMed]

30. Mehdi, H.; Naqvi, A.; Kamboh, M.I. A hydrophobic sequence at position 313–316 (Leu-Ala-Phe-Trp) in the fifth domain of apolipoprotein H (beta2-glycoprotein I) is crucial for cardiolipin binding. *Eur. J. Biochem.* **2000**, *267*, 1770–1776. [CrossRef] [PubMed]

31. Pennings, M.T.; Derksen, R.H.; van Lummel, M.; Adelmeijer, J.; VanHoorelbeke, K.; Urbanus, R.T.; Lisman, T.; de Groot, P.G. Platelet adhesion to dimeric beta-glycoprotein I under conditions of flow is mediated by at least two receptors: Glycoprotein Ibalpha and apolipoprotein E receptor 2'. *J. Thromb. Haemost.* **2007**, *5*, 369–377. [CrossRef] [PubMed]

32. Sheng, Y.; Sali, A.; Herzog, H.; Lahnstein, J.; Krilis, S.A. Site-directed mutagenesis of recombinant human beta 2-glycoprotein I identifies a cluster of lysine residues that are critical for phospholipid binding and anti-cardiolipin antibody activity. *J. Immunol.* **1996**, *157*, 3744–3751. [PubMed]

33. Van Lummel, M.; Pennings, M.T.; Derksen, R.H.; Urbanus, R.T.; Lutters, B.C.; Kaldenhoven, N.; de Groot, P.G. The binding site in {beta}2-glycoprotein I for ApoER2' on platelets is located in domain V. *J. Biol. Chem.* **2005**, *280*, 36729–36736. [CrossRef] [PubMed]

34. Lee, C.J.; De Biasio, A.; Beglova, N. Mode of interaction between beta2GPI and lipoprotein receptors suggests mutually exclusive binding of beta2GPI to the receptors and anionic phospholipids. *Structure* **2010**, *18*, 366–376. [CrossRef] [PubMed]

35. Amengual, O.; Atsumi, T.; Khamashta, M.A. Tissue factor in antiphospholipid syndrome: Shifting the focus from coagulation to endothelium. *Rheumatol. Oxf.* **2003**, *42*, 1029–1031. [CrossRef] [PubMed]

36. Kinev, A.V.; Roubey, R.A. Tissue factor in the antiphospholipid syndrome. *Lupus* **2008**, *17*, 952–958. [CrossRef] [PubMed]

37. Boles, J.; Mackman, N. Role of tissue factor in thrombosis in antiphospholipid antibody syndrome. *Lupus* **2010**, *19*, 370–378. [CrossRef] [PubMed]

38. Cuadrado, M.J.; Lopez-Pedrera, C.; Khamashta, M.A.; Camps, M.T.; Tinahones, F.; Torres, A.; Hughes, G.R.; Velasco, F. Thrombosis in primary antiphospholipid syndrome: A pivotal role for monocyte tissue factor expression. *Arthritis. Rheum.* **1997**, *40*, 834–841. [CrossRef] [PubMed]

39. Dobado-Berrios, P.M.; Lopez-Pedrera, C.; Velasco, F.; Aguirre, M.A.; Torres, A.; Cuadrado, M.J. Increased levels of tissue factor mrna in mononuclear blood cells of patients with primary antiphospholipid syndrome. *Thromb. Haemost.* **1999**, *82*, 1578–1582. [PubMed]

40. Nojima, J.; Masuda, Y.; Iwatani, Y.; Suehisa, E.; Futsukaichi, Y.; Kuratsune, H.; Watanabe, Y.; Takano, T.; Hidaka, Y.; Kanakura, Y. Tissue factor expression on monocytes induced by anti-phospholipid antibodies as a strong risk factor for thromboembolic complications in sle patients. *Biochem. Biophys. Res. Commun.* **2008**, *365*, 195–200. [CrossRef] [PubMed]

41. Lambrianides, A.; Carroll, C.J.; Pierangeli, S.S.; Pericleous, C.; Branch, W.; Rice, J.; Latchman, D.S.; Townsend, P.; Isenberg, D.A.; Rahman, A.; et al. Effects of polyclonal IgG derived from patients with different clinical types of the antiphospholipid syndrome on monocyte signaling pathways. *J. Immunol.* **2010**, *184*, 6622–6628. [CrossRef] [PubMed]

42. Satta, N.; Kruithof, E.K.; Fickentscher, C.; Dunoyer-Geindre, S.; Boehlen, F.; Reber, G.; Burger, D.; de Moerloose, P. Toll-like receptor 2 mediates the activation of human monocytes and endothelial cells by antiphospholipid antibodies. *Blood* **2011**, *117*, 5523–5531. [CrossRef] [PubMed]

43. Xie, H.; Zhou, H.; Wang, H.; Chen, D.; Xia, L.; Wang, T.; Yan, J. Anti-beta(2)GPI/beta(2)GPI induced TF and TNF-alpha expression in monocytes involving both TLR4/MYD88 and TLR4/TRIF signaling pathways. *Mol. Immunol.* **2013**, *53*, 246–254. [CrossRef] [PubMed]

44. Sorice, M.; Longo, A.; Capozzi, A.; Garofalo, T.; Misasi, R.; Alessandri, C.; Conti, F.; Buttari, B.; Rigano, R.; Ortona, E.; et al. Anti-beta2-glycoprotein I antibodies induce monocyte release of tumor necrosis factor alpha and tissue factor by signal transduction pathways involving lipid rafts. *Arthritis. Rheum.* **2007**, *56*, 2687–2697. [CrossRef] [PubMed]

45. Bazzan, M.; Vaccarino, A.; Stella, S.; Bertero, M.T.; Carignola, R.; Montaruli, B.; Roccatello, D.; Shoenfeld, Y.; Piedmont, A.P.S.C. Thrombotic recurrences and bleeding events in APS vascular patients: A review from the literature and a comparison with the APS piedmont cohort. *Autoimmun. Rev.* **2013**, *12*, 826–831. [CrossRef] [PubMed]

46. Cervera, R.; Serrano, R.; Pons-Estel, G.J.; Ceberio-Hualde, L.; Shoenfeld, Y.; de Ramon, E.; Buonaiuto, V.; Jacobsen, S.; Zeher, M.M.; Tarr, T.; et al. Morbidity and mortality in the antiphospholipid syndrome during a 10-year period: A multicentre prospective study of 1000 patients. *Ann. Rheum. Dis.* **2015**, *74*, 1011–1018. [CrossRef] [PubMed]

47. Chighizola, C.B.; Ubiali, T.; Meroni, P.L. Treatment of thrombotic antiphospholipid syndrome: The rationale of current management-an insight into future approaches. *J. Immun. Res.* **2015**, *2015*, 951424. [CrossRef] [PubMed]

48. Pedrinaci, S.; Ruiz-Cabello, F.; Gomez, O.; Collado, A.; Garrido, F. Protein kinase c-mediated regulation of the expression of CD14 and CD11/CD18 in U937 cells. *Int. J. Cancer* **1990**, *45*, 294–298. [CrossRef] [PubMed]

49. Fan, X.; Krahling, S.; Smith, D.; Williamson, P.; Schlegel, R.A. Macrophage surface expression of annexins I and II in the phagocytosis of apoptotic lymphocytes. *Mol. Biol. Cell* **2004**, *15*, 2863–2872. [CrossRef] [PubMed]

50. Greene, C.M.; McElvaney, N.G.; O'Neill, S.J.; Taggart, C.C. Secretory leucoprotease inhibitor impairs toll-like receptor 2- and 4-mediated responses in monocytic cells. *Infect. Immun.* **2004**, *72*, 3684–3687. [CrossRef] [PubMed]

51. Okamoto, M.; Hirai, H.; Taniguchi, K.; Shimura, K.; Inaba, T.; Shimazaki, C.; Taniwaki, M.; Imanishi, J. Toll-like receptors (TLRs) are expressed by myeloid leukaemia cell lines, but fail to trigger differentiation in response to the respective TLR ligands. *Br. J. Haematol.* **2009**, *147*, 585–587. [CrossRef] [PubMed]

52. Yang, X.V.; Banerjee, Y.; Fernandez, J.A.; Deguchi, H.; Xu, X.; Mosnier, L.O.; Urbanus, R.T.; de Groot, P.G.; White-Adams, T.C.; McCarty, O.J.; et al. Activated protein C ligation of ApoER2 (Lrp8) causes DAB1-dependent signaling in U937 cells. *Proc. Natl. Acad. Sci. USA* **2009**, *106*, 274–279. [CrossRef] [PubMed]

53. Kolyada, A.; Karageorgos, I.; Mahlawat, P.; Beglova, N. An A1-A1 mutant with improved binding and inhibition of beta2GPI/antibody complexes in antiphospholipid syndrome. *FEBS J.* **2015**, *282*, 864–873. [CrossRef] [PubMed]

54. Pennings, M.T.; Derksen, R.H.; Urbanus, R.T.; Tekelenburg, W.L.; Hemrika, W.; de Groot, P.G. Platelets express three different splice variants of ApoER2 that are all involved in signaling. *J. Thromb. Haemost.* **2007**, *5*, 1538–1544. [CrossRef] [PubMed]

55. Urbanus, R.T.; Pennings, M.T.; Derksen, R.H.; de Groot, P.G. Platelet activation by dimeric beta(2)-glycoprotein I requires signaling via both glycoprotein Ibalpha and apolipoprotein E receptor 2'. *J. Thromb. Haemost.* **2008**, *6*, 1405–1412. [CrossRef] [PubMed]

56. Agar, C.; van Os, G.M.; Morgelin, M.; Sprenger, R.R.; Marquart, J.A.; Urbanus, R.T.; Derksen, R.H.; Meijers, J.C.; de Groot, P.G. {beta}2-glycoprotein I can exist in two conformations: Implications for our understanding of the antiphospholipid syndrome. *Blood* **2010**, *116*, 1336–1343. [CrossRef] [PubMed]

57. Roubey, R.A.; Eisenberg, R.A.; Harper, M.F.; Winfield, J.B. "Anticardiolipin" autoantibodies recognize beta 2-glycoprotein I in the absence of phospholipid. Importance of Ag density and bivalent binding. *J. Immunol.* **1995**, *154*, 954–960. [PubMed]

58. Tincani, A.; Spatola, L.; Prati, E.; Allegri, F.; Ferremi, P.; Cattaneo, R.; Meroni, P.; Balestrieri, G. The anti-beta2-glycoprotein I activity in human anti-phospholipid syndrome sera is due to monoreactive low-affinity autoantibodies directed to epitopes located on native beta2-glycoprotein I and preserved during species' evolution. *J. Immunol.* **1996**, *157*, 5732–5738. [PubMed]

59. Willems, G.M.; Janssen, M.P.; Pelsers, M.M.; Comfurius, P.; Galli, M.; Zwaal, R.F.; Bevers, E.M. Role of divalency in the high-affinity binding of anticardiolipin antibody-beta 2-glycoprotein I complexes to lipid membranes. *Biochemistry* **1996**, *35*, 13833–13842. [CrossRef] [PubMed]

60. Pike, L.J. The challenge of lipid rafts. *J. Lipid Res.* **2009**, *50*, S323–S328. [CrossRef] [PubMed]

61. Lopez-Pedrera, C.; Buendia, P.; Cuadrado, M.J.; Siendones, E.; Aguirre, M.A.; Barbarroja, N.; Montiel-Duarte, C.; Torres, A.; Khamashta, M.; Velasco, F. Antiphospholipid antibodies from patients with the antiphospholipid syndrome induce monocyte tissue factor expression through the simultaneous activation of NF-kappab/Rel proteins via the p38 mitogen-activated protein kinase pathway, and of the MEK-1/ERK pathway. *Arthritis. Rheum.* **2006**, *54*, 301–311. [PubMed]

62. Zhou, H.; Wolberg, A.S.; Roubey, R.A. Characterization of monocyte tissue factor activity induced by IgG antiphospholipid antibodies and inhibition by dilazep. *Blood* **2004**, *104*, 2353–2358. [CrossRef] [PubMed]

63. Doring, Y.; Hurst, J.; Lorenz, M.; Prinz, N.; Clemens, N.; Drechsler, M.D.; Bauer, S.; Chapman, J.; Shoenfeld, Y.; Blank, M.; et al. Human antiphospholipid antibodies induce TNFalpha in monocytes via toll-like receptor 8. *Immunobiology* **2010**, *215*, 230–241. [CrossRef] [PubMed]

64. Kolyada, A.; Lee, C.J.; De Biasio, A.; Beglova, N. A novel dimeric inhibitor targeting beta2GPI in beta2GPI/antibody complexes implicated in antiphospholipid syndrome. *PLoS ONE* **2010**, *5*, e15345. [CrossRef] [PubMed]

Article

Potential Roles of Antiphospholipid Antibodies in Generating Platelet-C4d in Systemic Lupus Erythematosus

Chau-Ching Liu *, Travis Schofield, Amy Tang, Susan Manzi and Joseph M. Ahearn

Lupus Center of Excellence, Autoimmunity Institute, Allegheny Health Network Research Institute, Allegheny Health Network, Pittsburgh, PA 15212, USA; txs101TSS@aol.com (T.S.); amy.tang@ahn.org (A.T.); susan.manzi@ahn.org (S.M.); joseph.ahearn@ahn.org (J.M.A.)
* Correspondence: chauching.liu@ahn.org; Tel.: +1-412-359-6289

Academic Editor: Ricard Cervera
Received: 26 December 2016; Accepted: 21 June 2017; Published: 2 July 2017

Abstract: Premature, accelerated onset of atherothrombotic disease is prevalent in patients with systemic lupus erythematosus (SLE). Most, if not all, atherothrombotic diseases are likely to involve platelets and complement. Previously, we discovered that platelets bearing complement activation product C4d (P-C4d) are present in SLE patients, and are significantly associated with antiphospholipid (aPL) antibody positivity and stroke in SLE patients. The goal of the present study was to further elucidate the role of aPL and other platelet-reactive autoantibodies in the generation of P-C4d. To determine the association between P-C4d and aPL antibodies, the serum levels of aPL antibodies and P-C4d of 180 SLE patients were measured by enzyme-linked immunoassays and flow cytometry, respectively. To investigate the role of aPL antibodies, and possibly other autoantibodies as well, in mediating the generation of P-C4d, in vitro 2-step P-C4d induction experiments were performed. The results showed that the presence and levels of aPL antibodies in the serum were specifically elevated in SLE patients with positive P-C4d. The plasma and immunoglobulins purified from SLE patients who were positive for P-C4d and aPL were capable of inducing C4d deposition on normal platelets in vitro. The capacity of SLE plasma in inducing P-C4d appeared to correlate proportionately to the serum aPL levels. Collectively, the results demonstrate that both aPL and other platelet-reactive autoantibodies may participate in mediating the generation of P-C4d in SLE patients.

Keywords: antiphospholipid antibodies; anti-cardiolipin antibodies; anti-β_2 glycoprotein I antibodies; complement; platelet; systemic lupus erythematosus; cell-bound complement activation products (CB-CAPs)

1. Introduction

Premature, accelerated onset of atherosclerosis and thrombotic disease is prevalent in patients with systemic lupus erythematosus (SLE), a systemic autoimmune inflammatory disease characterized by autoantibody production, complement activation, and a myriad of clinical manifestations [1–4]. A continuously expanding spectrum of autoantibodies, including antiphospholipid antibodies, has been identified in SLE [5]. Antiphospholipid (aPL) antibodies encompass a heterogeneous group of antibodies with reactivity to various anionic phospholipid-binding proteins, including, primarily, anti-cardiolipin (aCL) antibodies, anti-β_2 glycoprotein I (aβ_2GPI) antibodies, and lupus anticoagulants [6–10]. In SLE, 30–40% of patients are found to be positive for aPL antibodies at some point of the disease course [11]. aPL antibodies have been shown to be associated with clinical events such as arterial and venous thrombosis and pregnancy complications in patients with antiphospholipid syndrome (APS) or SLE-associated APS (SLE/APS) [12–16].

Although the pathogenic roles of aPL antibodies in promoting thrombosis are not fully understood, abundant evidence suggests that these antibodies may function through disruption of the anticoagulant shield on cell surfaces, increase of oxidative stress, and activation of cells such as endothelial cells, monocytes, and platelets [17–22]. For example, aPL antibodies prepared from patients with APS or SLE/APS have been shown to bind to and activate platelets in vitro and in animals [23–28]. aPL antibodies may also trigger thrombosis and tissue injury via activation of the complement system [29]. In vivo studies using murine models have implicated the activation of the classical complement pathways in thrombosis and fetal loss associated with APS [30–32]. In humans, increased levels of complement activation products have been demonstrated in sera of patients with aPL antibodies who developed ischemic strokes [33].

We previously reported the presence of complement activation product C4d on the surface of platelets (platelet bound-C4d; P-C4d) in 18% of SLE patients, and identified a significant association between P-C4d positivity, aPL antibody positivity [34], and history of neurological manifestations in SLE patients (unpublished observations). In a recent study, P-C4d positivity was found to be significantly associated with stroke and all-cause mortality in SLE patients [35]. Other investigators have also identified associations between increased C4d deposition on platelets, and both arterial and venous thrombotic complications in patients with SLE [36–38]. In addition, P-C4d positivity was detected in 10% of patients with acute ischemic stroke without evidence of autoimmune disease, and correlated with stroke severity [39]. These observations, taken together, suggest an intriguing link between platelets, complement, aPL antibodies, and thrombotic disease in SLE.

Based on the studies outlined above, we hypothesize that in situ autoantibody (e.g., aPL)-mediated activation of the complement system generates C4d that can bind to platelet surfaces, and predispose platelets to a pro-thrombotic and pro-coagulating state, thereby promoting the development of thrombotic complications in patients with SLE. As a first step in verifying this hypothesis, we conducted an in-depth analysis of the prevalence and correlation of P-C4d and aPL antibodies in a cross-sectional study of 180 SLE patients, and investigated the role of aPL antibodies and other potential platelet-reactive autoantibodies in mediating C4d deposition on platelets in vitro.

2. Materials and Methods

2.1. Study Participants and Blood Specimens

All study participants were 18 years of age or older and provided written informed consent that was approved by the institutional review board. One hundred and eighty patients who met the American College of Rheumatology (ACR) 1982 or 1997 revised classification criteria for definite SLE ($\geq$4 criteria) [40,41] were recruited for this study during routine visits to the outpatient clinic of the Lupus Center of Excellence of the Allegheny Health Network, from July 2011 through June 2014. Two patients, who were hospitalized for disease flare and subsequently followed up at the outpatient clinic, were studied serially (up to December 2016). In addition, healthy volunteers were recruited to donate blood samples for platelet and serum preparation.

2.2. Plasma, Serum, and Immunoglobulin Preparation

At the time of each participant's visit, blood was collected into a Vacutainer tube containing ethylenediaminetetraacetic acid (EDTA) as an anticoagulant (for preparation of plasma and platelets) or a Vacutainer without anticoagulant (for serum preparation) (Becton Dickinson, Franklin Lakes, NJ, USA), and processed within 2 h after collection. An aliquot of the EDTA-anticoagulated whole blood was immediately used for P-C4d measurement. Plasma and sera were fractionated by centrifugation at $1600\times g$ for 10 min and stored at 4 °C (for immediate use) or -80 °C (for later use). Patient sera were used for measuring aPL antibodies.

Normal human serum prepared from the blood of healthy volunteers was aliquoted and stored at −80 °C for use in the in vitro P-C4d induction assays (see below). Immunoglobulin G (IgG) present in the plasma was isolated using a Pierce ImmunoPure® (A/G) IgG purification kit (Thermo Scientific, Rockford, IL, USA) following the manufacturer's instruction. After collection of IgG, the flow-through from the protein A/G affinity column was further fractionated to enrich for IgM using the Pierce Nab™ Protein L Spin purification kit (Thermo Scientific), following the manufacturer's instruction. The IgG and IgM fractions eluted from the respective affinity column were desalted, buffer-changed into phosphate-buffered saline (PBS), and concentrated by centrifugation using Microcon® centrifugal filters (molecular weight cutoff 30 kD; EMD Millipore, Billerica, MA, USA). To deplete Ig, the plasma sample was passed sequentially through the protein A/G column and protein L column twice. The final flow-through was collected, dialyzed against PBS, and concentrated back to the original volume using the Microcon® device. To deplete platelet-reactive autoantibodies, the SLE plasma sample (50 µL) was incubated with platelets (approximately 10^9) isolated from healthy volunteers at 4 °C for 30 min, and recovered by removal of platelets by centrifugation.

2.3. Flow Cytometry of P-C4d Measurement

C4d deposited on the surface of platelets (P-C4d) was measured by immunofluorescence staining/flow cytometry, as previously described [34]. Briefly, EDTA-anticoagulated whole blood was diluted in PBS prior to analysis. Platelets were electronically gated by forward scatter properties and expression of a platelet-specific marker, cluster of differentiation marker 42b (CD42b). Amounts of C4d present on platelets were assessed by using a monoclonal anti-C4d antibody (Quidel, San Diego, CA, USA) labeled with Alexa Fluor-488 using a Zenon mouse IgG1 labeling kit (Molecular Probes, Eugene, OR, USA). After staining, cells were analyzed using a FACS Calibur™ flow cytometer and Cell Quest Pro software (Becton Dickinson Immunocytometry Systems, San Jose, CA, USA). To ensure the specificity of P-C4d detected, blood aliquots from each patient stained with mouse IgG1 isotype were routinely included in all experiments. All monoclonal antibodies and Ig isotype control were used at a concentration of 5 µg/mL. Levels of P-C4d were expressed as specific median fluorescence intensity (SMFI), which was calculated as the C4d-specific median fluorescence intensity minus the isotype control median fluorescence intensity. P-C4d was considered positive based on the cut-off point of 2.15 as previously reported [34], which was derived from repeated measures of P-C4d in 100 healthy individuals.

2.4. aPL Antibody Immunoassays

Serum levels of aPL antibodies were determined in blood samples collected at the same time for P-C4d measurement. Serum samples of individual patients were prepared within 2 h of blood collection and stored at −80 °C until analysis. Levels of isotype-specific anti-cardiolipin (aCL) and anti-β_2 glycoprotein I (aβ_2GPI) antibodies in the serum were determined using the EL-aCL™ (IgM-IgG-IgA) ELISA kit and EL-β_2GPI™ (IgM-IgG-IgA) ELISA kit (TheraTest Labs, Lombard, IL, USA), respectively, following the manufacturer's instruction. Results are shown as standard MPL, GPL, and APL units. Positive levels for all aPL antibodies were defined following normalization to appropriate assay calibrators supplied by the manufacturer. The presence or absence of lupus anticoagulants was not assessed at the time of study.

2.5. In Vitro P-C4d Induction Assay

The capacity of autoantibodies in the plasma of SLE patients to active the complement system and induce C4d deposition of platelets was assessed in vitro. Briefly, platelet-rich plasma was prepared by centrifugation of EDTA-anticoagulated blood obtained from healthy volunteers or P-C4d-negative SLE patients, at 120× *g* for 10 min. Platelets were then collected and washed twice with PBS by centrifugation at 1600× *g*, in the presence of 1 g/mL of prostaglandin E1 (Sigma Aldrich, St. Louis, MO, USA). The resulting platelets were fixed with 0.5% paraformaldehyde in PBS for 20 min, washed once with PBS, and resuspended in PBS. Aliquots of platelet suspension were incubated with plasma samples (at 20% final concentration) prepared from selected SLE patients who had been identified as having positive P-C4d or high aPL serum levels. In some experiments, the SLE plasma was replaced with different amounts of purified IgG or IgM, Ig-depleted plasma, platelet-preabsorbed plasma, or commercially available human aCL antibody (Immunovision). After incubation at 4 °C for 45 min to allow autoantibody binding, the platelet samples were washed with PBS twice, resuspended in 50 μL of GVB^{2+} buffer (1% gelatin, 5 mM Na veronal, 142 mM NaCl, pH 7.3, containing Ca^{2+} and Mg^{2+}), and incubated with 10 mL of normal human serum (as a source of complement). After incubation at 37 °C for 1 h to allow complement activation, the platelet suspensions were washed with PBS twice and subjected to P-C4d measurement by flow cytometry, as described above.

2.6. Statistical Analysis

Data were presented as mean and standard deviation (SD), or median and interquartile range (IQR), for continuous variables based on their distributions, and as frequency and percentage for categorical variables. Baseline comparisons of continuous variables between SLE patients with or without aPL antibodies were performed using two-sample *t* or Wilcoxon rank-sum (Mann–Whitney) tests, as appropriate. Categorical variables were analyzed using or Chi-Square or Fisher's Exact tests. All *p*-values were considered significant at $p < 0.05$. All analyses were performed using the STATA/SE version 11.0 for Windows (Stata Corporation, College Station, TX, USA). Nonparametric technique (permutation test) was used to perform analysis of one-way multivariate data (IgA, IgM, and IgA) with approximations for ANOVA Type, Wilks' Lambda, Lawley Hotelling, and Bartlett–Nanda–Pillai Test statistics. Multiple testing algorithms were used to control the familywise error rate [42].

3. Results

3.1. Patient Characteristics

The demographics and clinical characteristics of the 180 patients with SLE are shown in Table 1. Thirty-four patients (18.9%) were found to be P-C4d-positive (P-C4d levels >2.15) at the study visit, consistent with our previous findings [34,35]. Compared to P-C4d-negative patients (*n* = 146), P-C4d-positive patients are more likely to have a history of hematologic involvement (hemolytic anemia, lymphopenia, and thrombocytopenia) and aPL positivity (aCL and lupus anticoagulants). A significantly higher percentage of P-C4d-positive patients were on anticoagulant therapy than were P-C4d-negative patients (warfarin, *p* = 0.005; heparin, *p* = 0.025), suggesting that P-C4d-positive patients were more likely to have experienced, or were considered to be at increased risk for, thrombotic complications. An increased fraction of P-C4d-positive patients (73.5% vs. 57.5% of P-C4d-negative patients; *p* = 0.074) received steroid treatment at the time of visit, suggesting possibly higher disease activity in these patients.

Table 1. Patient demographics and clinical characteristics.

		Platelet-C4d		
	All (*n* = 180)	Positive (*n* = 34)	Negative (*n* = 146)	*p*-Value
Age, year, mean (SD)	47.2 (11.9)	46.9 (10.7)	47.3 (12.2)	0.802
Duration of SLE, year, mean (SD)	15.2 (9.7)	16.5 (9.6)	14.8 (9.7)	0.276
Sex, *n* (%)				
Women	167 (92.8)	31 (91.2)	136 (93.2)	0.714
Race, *n* (%)				
White	156 (86.7)	28 (82.4)	128 (87.7)	0.240
Black	21 (11.7)	6 (18.6)	15 (10.3)	
Others	3 (1.7)		3 (2.0)	

Table 1. *Cont.*

	Platelet-C4d			
	All (*n* = 180)	Positive (*n* = 34)	Negative (*n* = 146)	*p*-Value
ACR [&] criteria (ever), *n* (%)				
Malar rash	83 (46.1)	13 (38.2)	70 (47.9)	0.344
Discoid rash	14 (7.8)	4 (11.8)	10 (6.8)	0.306
Photosensitivity	111 (61.7)	17 (50.0)	94 (64.4)	0.170
Oral ulcers	122 (67.8)	19 (55.9)	103 (70.3)	0.107
Arthritis	172 (95.6)	32 (94.1)	140 (95.9)	0.647
Serositis	64 (35.6)	14 (41.2)	50 (34.2)	0.551
Renal disease	54 (30.0)	8 (23.5)	46 (31.5)	0.412
Neurological	20 (11.1)	4 (11.8)	16 (10.9)	1.000
Seizure	15 (8.3)	1 (2.9)	14 (9.6)	0.310
Psychosis	9 (5.0)	3 (8.8)	6 (4.1)	0.402
Hematological	96 (53.3)	25 (73.5)	71 (48.6)	0.012
Hemolytic anemia	10 (5.6)	6 (17.6)	4 (2.7)	0.004
Leukopenia	46 (25.6)	11 (32.4)	35 (24.0)	0.382
Lymphopenia	61 (33.9)	19 (55.9)	42 (22.8)	0.004
Thrombocytopenia	32 (17.8)	10 (29.4)	22 (15.1)	0.047
Antinuclear antibody	177 (98.3)	33 (97.1)	144 (98.6)	0.468
Serological	142 (78.9)	33 (97.1)	109 (74.7)	0.002
Anti-Phospholipid [#,^]	83 (46.1)	24 (70.6)	59 (40.4)	0.002
Anti-cardiolipin	60 (33.3)	17 (50.0)	43 (29.5)	0.028
Lupus anticoagulant	45 (25.0)	16 (47.1)	29 (19.9)	0.002
Anti-dsDNA	97 (53.9)	23 (67.6)	74 (50.7)	0.087
Anti-Smith	24 (13.3)	4 (11.8)	20 (13.7)	1.000
Medication use, current [**], *n* (%)				
Steroid	109 (60.6)	25 (73.5)	84 (57.5)	0.074
Antimalarial	128 (71.1)	23 (67.6)	105 (71.9)	0.676
Anticoagulant				
Warfarin	24 (13.3)	10 (29.4)	14 (9.6)	0.005
Heparin	7 (3.9)	4 (11.8)	3 (2.1)	0.025
FXa inhibitor	2 (1.1)	1 (2.9)	1 (0.7)	0.343
Antiplatelet				
Aspirin	69 (38.3)	11 (32.4)	58 (39.7)	0.557
P2Y12 antagonist	10 (5.6)	4 (11.8)	6 (4.1)	0.096
Statins	37 (20.6)	7 (20.6)	30 (20.5)	1.000
Immunosuppressant	101 (56.1)	21 (61.8)	80 (50.8)	0.566
Biologicals	18 (10.0)	4 (11.8)	14 (9.6)	0.751
Antihypertensive	77 (42.8)	13 (38.2)	64 (43.8)	0.571
Diuretics	31 (17.2)	3 (8.8)	28 (19.2)	0.208
Insulin	6 (3.3)	0 (0.0)	6 (4.1)	0.592
Antidiabetic	4 (2.2)	0 (0.0)	4 (2.7)	1.000
P-C4d level, current, median (IQR)	0.6 (0.2–1.4)	5.6 (3.4–11.1)	0.4 (0.1–0.8)	<0.001

* Continuous variables are presented as mean (SD) or median depending on the data distribution. [&] ACR: American College of Rheumatology. [#] Eighty-three patients had a history of aPL antibody positivity prior to the current study visit. [^] Of the 83 historically aPL-positive patients, 32 remained positive at the current study visit. In addition, 16 patients without a history of aPL antibody positivity were tested positive at the current study visit (see Table 2 for further information). [**] Medication use at the study visit: antimalarial—hydroxychloroquine, chloroquine, and quinacrine; immunosuppressants—azathioprine, methotrexate, mycophenolate mofetil, mycophenolic acid, cyclophosphamide, tacrolimus, and leflunomide; biologicals—belimumab. P-C4d: platelet-bound C4d; aPL: antiphospholipid. SLE: systemic lupus erythematosus. IQR: interquartile range.

3.2. Relationships between P-C4d and aPL Antibodies

In a previous cross-sectional study, P-C4d positivity was found to be significantly associated with aPL positivity (specifically, lupus anticoagulants, aCL IgG, and aCL IgM) [34]. The present study was aimed at further examining the prevalence of aPL antibodies and their correlation with P-C4d in SLE patients. Using commercial immunoassay kits, levels of isotype-specific aCL and aβ_2GPI antibodies in the serum samples obtained concomitantly with P-C4d measure were determined (Table 2). In the present SLE patient cohort, 26.7% (48/180) were positive for aPL (aCL and/or aβ$_2$GPI) antibodies at the study visit (Table 2). Notably, the P-C4d-positive subgroup had a considerably higher rate of aPL positivity than the P-C4d-negative subgroup (58.5% vs. 19.2%; $p < 0.001$). Compared to P-C4d-negative patients, P-C4d-positive patients were significantly more likely to have higher levels of aPL antibodies in 4 of the 6 isotypes studied (aCL IgG, aβ$_2$GPI IgG, aβ$_2$GPI IgM, and aβ$_2$GPI IgA). Although the positivity rates of aCL IgM and aCL IgA were significantly increased in P-C4d-positive patients than in P-C4d-negative patients (both $p = 0.012$), the serum levels of these two aPL antibodies were not significantly higher in P-C4d-positive patients ($p = 0.068$ and $p = 0.053$, respectively). Of the 48 patients positive for aPL (any aCL or aβ$_2$GPI isotype), 41.9% (20/48) had elevated levels of aPL antibodies of multiple isotypes (range: 2–6) at the study visit (Table 3). Notably, P-C4d-positive patients were more likely to have multiple isotypes of aPL antibodies than P-C4d-negative patients. For example, 8 out of 13 P-C4d-positive patients who had IgA aCL or aβ$_2$GPI also had IgG and/or IgM aCL or aβ$_2$GPI, whereas only 1 out of 8 P-C4d-negative patients who had IgA aCL or aβ$_2$GPI also had IgG aCL. This finding is consistent with the fact that IgA is incapable of inducing complement activation via the classical pathway, but the concomitant presence of IgG and/or IgM aPL antibodies may be responsible for inducing complement activation on platelets. Moreover, aPL-positive-patients had significantly higher levels of P-C4d than did aPL-negative-patients ($p = 0.002$). Among the aPL-positive-patients, the P-C4d levels were differentially elevated in patients with aCL antibodies alone or aβ$_2$GPI antibodies alone, and were markedly elevated in those positive for both aCL and aβ$_2$GPI antibodies (Table 4).

Table 2. Comparison of serum anti-Cardiolipin (aCL) and anti-β$_2$GPI (aβ$_2$GPI) levels in P-C4d-positive vs. P-C4d-negative patients.

	Platelet-C4d			
	All (n = 180)	Positive (n = 34)	Negative (n = 146)	p-Value (P-C4d+ vs. P-C4d−)
aPL (aCL and/or aβ$_2$GPI) antibodies positivity, n (%)	48 (26.7)	20 (58.5)	28 (19.2)	<0.001
aCL IgM positivity, n (%) [†]	19 (10.6)	8 (23.5)	11 (7.5)	0.012
aCL IgM level (U/mL) median (IQR) *	3.1 (2.0–6.5)	4.2 (2.2–11.6)	2.8 (2.0–6.0)	0.068
aCL IgG positivity, n (%) [†]	17 (9.4)	11 (32.4)	6 (4.1)	<0.001
aCL IgG level (U/mL) median (IQR) *	5.0 (2.8–9.6)	12.5 (5.6–37.9)	4.6 (2.7–7.7)	<0.001
aCL IgA positivity, n (%) [†]	6 (3.3)	4 (11.8)	2 (1.4)	0.012
aCL IgA level (U/mL) median (IQR) *	1.2 (0.6–2.3)	1.3 (1.0–2.9)	1.0 (0.5–2.0)	0.053
Any aCL positivity, n (%)	32 (17.8)	13 (38.2)	19 (13.0)	0.002
aβ$_2$GPI IgM positivity, n (%)	10 (5.6)	5 (14.7)	5 (3.4)	0.022
aβ$_2$GPI IgM level (U/mL) median (IQR)	0.4 (0.0–1.3)	1.0 (0.4–3.6)	0.3 (0.0–0.9)	<0.001
aβ$_2$GPI IgG positivity, n (%)	14 (7.8)	11 (32.4)	8 (5.5)	<0.001
aβ$_2$GPI IgG level (U/mL) median (IQR)	1.6 (0.0–3.7)	4.6 (1.1–76.1)	1.3 (0.0–3.1)	<0.001
aβ$_2$GPI IgA positivity, n (%)	21 (11.7)	13 (38.2)	16 (11.0)	<0.001
aβ$_2$GPI IgA level (U/mL) median (IQR)	0.4 (0.0–2.6)	3.9 (0.8–12.8)	0.4 (0.0–1.2)	<0.001
Any aβ$_2$GPI positivity, n (%)	34 (18.9)	18 (52.9)	16 (11.0)	<0.001

[†] Positivity for all aPL antibodies was defined based on the cutoff values provided by the ELISA kits used. * median (IQR). C4d−: C4d-negative; C4d+: C4d-positive; aβ$_2$GPI: anti-β$_2$ glycoprotein I.

Table 3. Distribution of anti-Cardiolipin (aCL) and anti-β_2GPI (aβ_2GPI) antibodies in SLE patients.

Number of Positive aCL and aβ_2GPI Antibodies	Number of Patients	Isotypes of Positive aCL and aβ_2GPI Antibodies
6	1	aCL IgM/IgG/IgA; aβ_2GPI IgM/IgG/IgA
5	1	aCL IgM/IgG/IgA; aβ_2GPI IgG/IgA
	1	aCL IgM/IgG/IgA; aβ_2GPI IgM/IgG
4	1	aCL IgG/IgA; aβ_2GPI IgG/IgA
	1	aCL IgM/IgG; aβ_2GPI IgG/IgA
3	2	aCL IgG; aβ_2GPI IgG/IgA
	1	aCL IgM/IgG; aβ_2GPI IgG
	1	aβ_2GPI IgM/IgG/IgA
	1	aCL IgM/IgG/IgA
2	3	aCL IgM; aβ_2GPI IgM
	5	aCL IgG; aβ_2GPI IgG
	2	aCL IgM; aβ_2GPI IgA
1	8	aCL IgM
	3	aCL IgG
	2	aCLIgA
	2	aβ_2GPI IgM
	1	aβ_2GPI IgG
	12	aβ_2GPI IgA

Table 4. Comparison of P-C4d levels in SLE patients with or without aPL antibodies.

	P-C4d Median (IQR) *	*p*-Value (vs. aPL−)
aPL− (*n* = 132)	0.52 (0.13–0.91)	
aPL+ (*n* = 48) **	1.18 (0.25–4.55)	0.002
aCL+ alone (*n* = 14)	0.48 (0.20–1.51)	0.586
aβ_2GPI+ alone (*n* = 16)	1.44 (0.25–3.24)	0.042
aCL+/aβ_2GPI+ (*n* = 18)	4.61 (0.53–11.44)	0.001

* P-C4d (specific median florescence intensity; SMFI): median (interquartile range). ** aPL+: antiphospholipid antibody positive including aCL and/or aβ_2GPI antibodies. aPL+: aPL-positive; aPL−: aPL-negative.

3.3. Involvement of aPL Antibodies in P-C4d Generation

The observed concomitant presence of abnormal levels of aPL antibodies and P-C4d in SLE patients, together with the literature reporting binding of aPL antibodies to platelets [26,27,43], suggests that aPL antibodies constitute a major category of platelet-reactive autoantibodies and are involved in the generation and deposition of C4d on platelets in SLE patients. Therefore, we investigated whether the plasma of SLE patients contains aPL antibodies (and perhaps other autoantibodies reactive to platelets) that can bind to and induce C4d deposition on platelets in vitro. To circumvent potential confounding effects resulting from hypocomplementemia and medication (e.g., heparin) in SLE samples, and to provide an equally sufficient amount of complement in all experiments, we designed a 2-step protocol. Platelets were first incubated with SLE plasma to allow binding of antibodies, washed, and then incubated with a fixed volume of normal human serum to allow for complement activation (Figure 1A). As represented in Figure 1B, platelets derived from healthy individuals acquired significant levels of C4d on their surface after being incubated with plasma prepared from SLE patients who were tested positive for aPL antibodies and had C4d deposited on platelets ex vivo (C4d+/aPL+). In contrast, platelets were not bound by C4d after being incubated with plasma prepared either from SLE patients who had no detectable aPL antibodies or C4d on platelets ex vivo (C4d−/aPL−), or from healthy controls. Interestingly, plasma samples prepared from SLE patients who were aPL-positive but had no detectable level of P-C4d (C4d−/aPL+) were found incapable of inducing C4d deposition in vitro. When a complement component 1q (C1q)-depleted human serum or heat-inactivated normal human serum was used as the source of complement in the assay, the capacity of SLE plasma to induce P-C4d was lost (Figure 1C). In our previous study, we did not find C4d deposition on platelets of patients with primary APS ex vivo [34]. Similarly, the plasma prepared from a representative patient with primary APS was unable to induce P-C4d in vitro (Figure 1D).

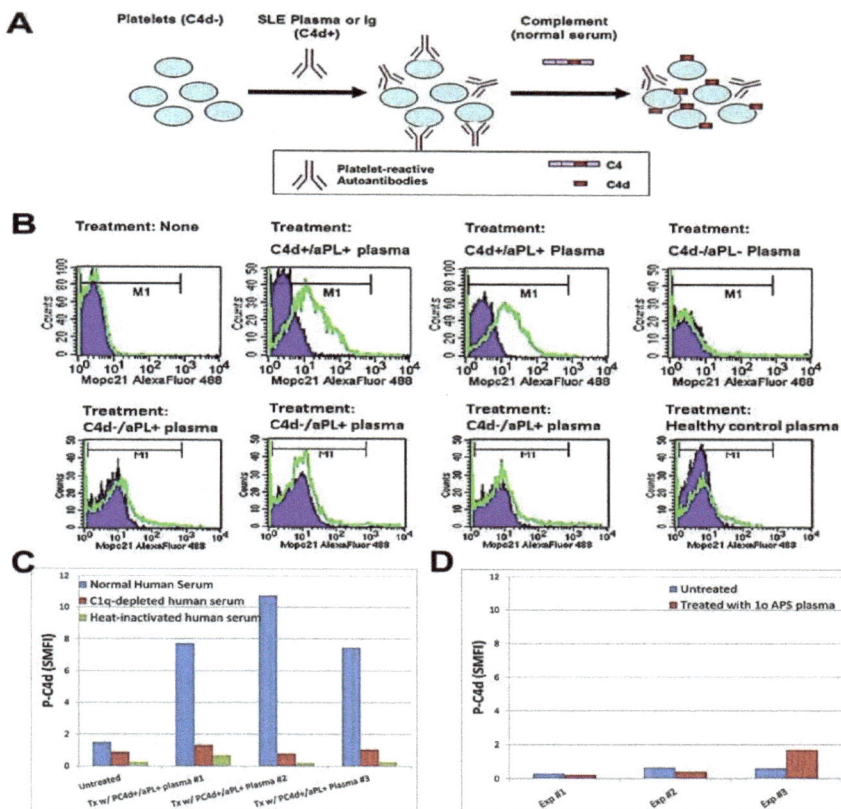

Figure 1. Induction of C4d deposition on the surface of platelets in vitro. (**A**) Schematic illustration of the in vitro P-C4d induction experiments. C4d-negative (C4d−) cells are incubated with plasma or Ig isolated from C4d-positive (C4d+) SLE patients, and subsequently with normal human serum (complement). After treatment, levels of C4d deposited on platelet surfaces were analyzed by flow cytometry; (**B**) Platelets from a healthy individual were incubated with the plasma of two P-C4d+/aPL+ SLE patients, one P-C4d−/aPL− SLE patient, three P-C4d−/aPL+ SLE patients, and one healthy control, followed by incubation with complement. Note the high levels of C4d deposited on cell surfaces after treatment with P-C4d+/aPL+ SLE plasma, but not with P-C4d−/aPL− SLE plasma, P-C4d−/aPL+ SLE plasma, or healthy control plasma. Purple histogram: isotype control; green open histogram: anti-C4d; (**C**) C4d deposition was induced in vitro only when platelets were treated with P-C4d+/aPL+ plasma, followed by normal human serum. C1q-depleted or heat-inactivated human serum was incapable of inducing C4d deposition; (**D**) Plasma of patients with primary antiphospholipid syndrome was incapable of inducing C4d deposition on platelets in vitro. P-C4d: platelet-bound C4d; aPL: antiphospholipid antibodies; P-C4d+/aPL+: P-C4d positive/aPL positive; P-C4d−/aPL−: P-C4d negetive /aPL negative; P-C4d−/aPL+: P-C4d negetive/aPL negative; SLE: systemic lupus erythematosus.

To further investigate the role of aPL antibodies in mediating P-C4d generation, plasma samples were serially collected from an SLE patient (#107395) whose serum aPL levels decreased over a 15-month period during 2012–2013. The capacity of these plasma samples to induce C4d deposition correlated with the aPL levels in the plasma (Figure 2A). Such correlation was noted again when she had a subsequent episode of SLE/APS flare in 2016 (Figure 2B). A similar observation was made in another SLE patient (#209310) (Figure 2C). Moreover, a commercially available human aCL IgG antibody (ImmunoVision, Inc., Springdale, AR, USA) was capable of inducing C4d deposition on platelets in a dose-dependent manner (Figure 2D). In comparison, human anti-ribosomal P (Figure 2E) and anti-dsDNA (Figure 2F) antibodies (ImmunoVision) did not induce P-C4d in vitro. Collectively, these results provide convincing support for a pivotal role of aPL antibodies in mediating generation and deposition of C4d on platelets.

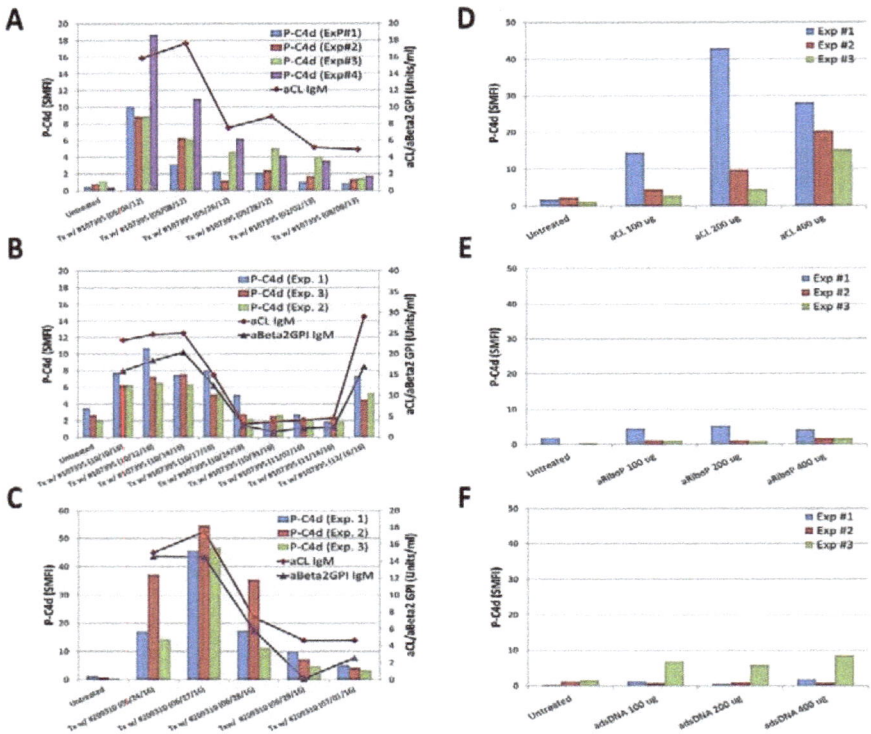

Figure 2. Participation of anti-phospholipid antibodies in generating the P-C4d phenotype in vitro. (**A–C**) Platelets prepared from healthy controls were untreated, or treated with serial plasma samples collected from an SLE patient (#107395) who was positive for aCL IgM and aβ₂GPI IgM antibodies during two longitudinal follow-up periods during 2012–2013 (panel A) and 2016 (panel B), respectively. Plasma samples serially collected from another SLE patient (#209310) who was positive for aCL IgM and aβ₂GPI IgM antibodies were similarly tested (panel C); (**D–F**) Platelets prepared from a healthy control or SLE patient were untreated or treated with a commercially available human aCL antibody (panel E), anti-ribosomal P antibody (panel E), or anti-double stranded DNA (dsDNA) antibody (panel F). Note that P-C4d was induced in a dose-dependent manner by aCL antibody, but not by anti-ribosomal P or anti-dsDNA antibody. Results shown are the mean and standard deviation derived from three independent experiments.

3.4. Potential Involvement of Other Platelet-Reactive Autoantibodies in P-C4d Generation

To date, we have tested and compared plasma samples derived from different SLE patients for their capacity to induce generation of P-C4d in vitro. Overall, the majority of aPL-positive plasma samples were able to induce P-C4d efficiently, and most aPL-negative plasma samples were unable to induce P-C4d (see Figure 1B and Table 5 for representative data). However, exceptional cases existed. In some cases, plasma samples derived from SLE patients who had elevated P-C4d but otherwise tested negative for aPL antibodies were found capable of inducing C4d deposition on platelets of some SLE patients in vitro (Table 5; SLE patient #4). This finding suggested that antibodies other than aPL may react to platelets and mediate complement activation in situ on the surface of platelets. We next performed in vitro P-C4d induction experiments using Ig purified from the plasma of SLE patients. The results demonstrated that Ig purified from the plasma of SLE patients with elevated P-C4d levels and aPL antibodies were capable of inducing P-C4d deposition on platelets in vitro, while Ig purified from SLE patients with negligible P-C4d and aPL antibodies could not induce P-C4d in vitro (Figure 3A). Moreover, depletion of Ig from SLE plasma completely abolished its capacity to induce complement activation and C4d deposition on platelets in vitro (Figure 3B). This capacity of the SLE plasma was significantly decreased if the potential platelet-reactive antibodies were removed by pre-incubation of the plasma with platelets (Figure 3B). This latter result indicates that antibodies reactive to platelets are responsible for mediating the generation of P-C4d.

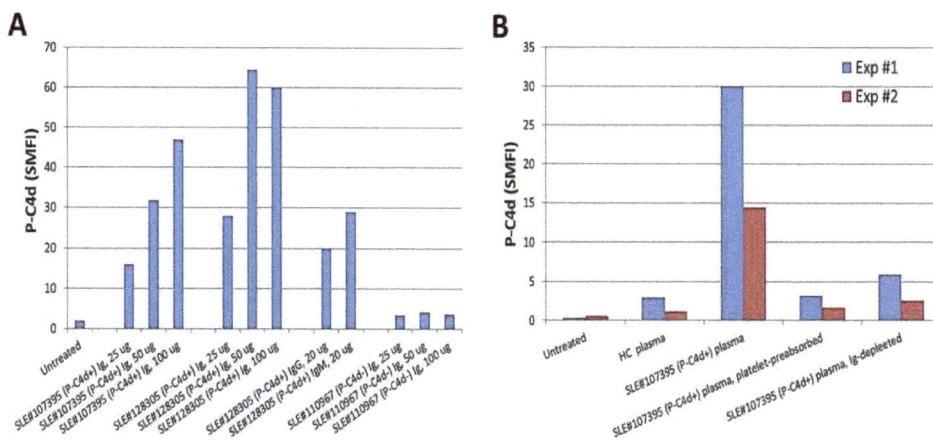

Figure 3. Presence of platelet-reactive autoantibodies in plasma of SLE patients. (**A**) Platelets from a healthy control were incubated with different amounts of immunoglobulins purified from two representative P-C4d+/aPL+ SLE patients (#107395 and #128305), and from a representative P-C4d−/aPL− SLE patient (#110967). P-C4d levels induced were determined by flow cytometric analysis. Note that Ig prepared from the P-C4d+/aPL+ patients induced P-C4d in a dose-dependent manner; (**B**) Platelets derived from two healthy individuals (Exp#1 and Exp#2) were treated with plasma samples derived from a healthy control (HC) or from a representative P-C4d+/aPL+ SLE patient (#107395). The SLE plasma sample was further depleted of platelet-reactive autoantibodies by pre-absorption with platelets, or depleted of immunoglobulins (Ig) using protein A/G/L.

Table 5. Comparison of the capacity of aPL+ and aPL− SLE plasma to induce P-C4d in vitro.

Plasma Source [a]	Plasma aPL Positivity [b]	Ex vivo P-C4d Positivity [c]	Platelet Source [d]	In Vitro P-C4d (SMFI) [e]
SLE patient #1	+	+	SLE patient #4	34.21
SLE patient #1			SLE patient #5	39.17
SLE patient #1			SLE patient #6	19.31
SLE patient #1			SLE patient #7	14.70
SLE patient #1			SLE patient #8	6.83
SLE patient #1			SLE patient #9	27.34
SLE patient #1			SLE patient #10	39.29
SLE patient #1			SLE patient #11	31.52
SLE patient #1			Healthy Control #1	27.50
SLE patient #1			Healthy Control #2	5.38
SLE patient #2	+	+	SLE patient #4	4.93
SLE patient #2			SLE patient #7	5.20
SLE patient #2			SLE patient #8	0.44
SLE patient #2			SLE patient #9	6.20
SLE patient #2			SLE patient #10	9.37
SLE patient #2			SLE patient #11	9.53
SLE patient #2			SLE patient #12	3.36
SLE patient #2			Healthy Control #1	1.15
SLE patient #2			Healthy Control #2	0.78
SLE patient #3	+	+	SLE patient #12	20.48
SLE patient #3			Healthy Control #1	24.92
SLE patient #3			Healthy Control #2	14.30
SLE patient #4	−	+	SLE patient #5	3.07
SLE patient #4			SLE patient #6	0.80
SLE patient #4			SLE patient #7	2.08
SLE patient #4			SLE patient #8	0.26
SLE patient #4			SLE patient #9	3.96
SLE patient #4			SLE patient #10	7.20
SLE patient #4			SLE patient #11	8.70
SLE patient #4			Healthy Control #1	0.10
SLE patient #4			Healthy Control #2	1.14

[a] Plasma samples used in the first step of the in vitro P-C4d induction experiments. [b] Positivity for aCL and/or aβ_2GPI antibodies in the plasma source determined by ELISA. [d] Positivity of P-C4d on platelets of the donor of the plasma source. [d] Platelets derived from healthy controls or SLE patients who have no/low C4d on platelet surfaces. [e] C4d levels on the surface of treated platelets, determined by flow cytometry. Boldfaced numbers indicate positive P-C4d (SMFI. 2.15). +: positivity; −: negativity.

4. Discussion

We have previously demonstrated that complement activation products, particularly C4d, bind at high levels to circulating cells in patients with SLE, thereby generating a cell-bound complement activation product (CB-CAP) signature that is highly sensitive and specific for a lupus diagnosis [44,45]. Until recently, the mechanism(s) responsible for generating the CB-CAP signature have not been systematically investigated. Our recent study has shown that anti-lymphocyte autoantibodies play a pivotal role in generating patient-specific T-cell-bound C4d (T-C4d) signatures [46], providing solid evidence for an autoantibody-mediated mechanism underlying CB-CAP generation for the first time. Of the various CB-CAP phenotypes identified to date, platelet-bound C4d (P-C4d) is characteristically distinct in that it is identified only in a relatively small fraction of SLE patients (approximately 20% in sensitivity), yet in an extremely exclusive manner (99% in specificity) [34]. P-C4d was shown to be significantly associated with all-cause mortality and ischemic stroke in SLE patients [35]. Other investigators using a similar but experimentally distinct approach have observed that deposition of complement proteins C1q, C4d, and C3d are increased on platelets of patients with SLE (~48% sensitivity), and to a lesser extent, on platelets of patients with other autoimmune diseases [37,38]. Interestingly, those investigators also reported that increased C4d deposition on platelets is associated with vascular events in patients with SLE [36–38]. Collectively, these studies support a pathogenic role of P-C4d in a subset of SLE patients at increased risk for thrombotic events.

Given the unique features of P-C4d, the present study is focused specifically on the mechanism underlying P-C4d generation. Here we provide several lines of evidence supportive of the role of aPL antibodies and platelet-reactive autoantibodies in the generation of P-C4d. These include: (1) P-C4d-positive SLE patients had not only a higher frequency but also significantly elevated serum levels of aCL and $a\beta_2$GPI antibodies; (2) plasma of SLE patients was capable of inducing C4d deposition on platelets, in vitro, in an aCL/$a\beta_2$GPI concentration-dependent manner; (3) purified aCL antibody and Ig derived from SLE patients with aPL antibodies could induce C4d deposition on platelets in vitro; (4) plasma prepared from some SLE patients without aPL antibodies was also capable of inducing C4d deposition on platelets in vitro; and (5) this capacity was abolished by pre-absorption of the plasma with platelets.

In the inaugural study of P-C4d as a biomarker for SLE [34], it was reported that P-C4d was significantly associated with positivity for lupus anticoagulants and IgG/IgM aCL antibodies. This initial observation prompted us to speculate that aPL antibodies may be important participants in mediating the P-C4d phenotype in SLE patients. We sought to further elucidate the role of aPL antibodies through the present cross-sectional study, in which P-C4d and aPL antibody levels of a given patient were measured on the same study date, and plasma samples with known aPL antibody concentrations were used for in vitro P-C4d induction experiments. Because the presence and levels of aPL antibodies tend to wax and wane over time, this approach allows for a more definite investigation of aPL antibodies, compared to the use of historical data to indicate the aPL status of a given patient sample tested. Indeed, one patient in our SLE cohort was treated twice for a disease flare and followed longitudinally over a period of four years (2012–2016). In each flare episode, her aPL antibody levels decreased following treatment and, in parallel, the capacity of her plasma to induce P-C4d generation in vitro decreased (Figure 2A,B). Similar observations were made in another patient followed in a prospective manner (Figure 2C). The 2-step in vitro P-C4d induction assay used in the present study also allows us to decipher the complement pathway mediated by aPL antibodies. When a C1q-depleted human serum or heat-inactivated normal human serum was used as the source of complement in the assay, the capacity of SLE plasma to induce P-C4d was lost (Figure 1C). Taken together, these results provide convincing support for the involvement of aPL antibodies in mediating the generation and deposition of C4d on platelets through activation of the complement classical pathway.

In the present study, 28 SLE patients were determined to be aPL-positive yet P-C4d-negative at the study visit. Plasma derived from these aPL-positive/P-C4d-negative patients was unable to induce C4d deposition on platelets in vitro (Figure 1B). Compared to SLE patients who were positive for both aPL and P-C4d, aPL-positive/P-C4d-negative SLE patients appeared to have lower (albeit within the "positive" range) levels of aPL antibodies, and were less likely to test positive for multiple isotypes of aPL antibodies. Six of the aPL-positive/P-C4d-negative patients had only the IgA isotype of $a\beta_2$GPI antibody, and 2 patients had only the IgA isotype of aCL antibody, an isotype incapable of inducing complement activation via the classical pathway. Five and thirteen patients were positive for aCL IgG and $a\beta_2$GPI IgG, respectively. It is possible that IgG aPL antibodies in those patients are of the IgG2 or IgG4 subclasses that are inefficient in triggering complement activation.

The discoveries reported here support and extend those made recently by others. Lood and colleagues have demonstrated that increased levels of complement activation products, including C1q, C3d, and C4d, were detected on the surface of platelets of SLE patients, especially those with a history of venous thrombosis [37]. Using a one-step in vitro assay that is distinct from our current assay, these investigators also found that serum of SLE patients with a history of lupus anticoagulant supported deposition of C4d on platelets. In a more recent study, Lood et al. further reported that SLE patients with aPL antibodies had higher P-C4d levels than those without aPL antibodies, and that aPL antibodies can activate platelets and induce C4d deposition in vitro [38]. Peerschke and colleagues demonstrated that sera from SLE patients could fix complement and induce C1q/C4d deposition on platelets in vitro, measured by a solid phase-based enzyme-linked immunosorbent assay [36]; this complement fixation/activation activity was associated with the presence of IgG aCL and $a\beta_2GPI$ antibodies in serum samples tested.

In the current study cohort, 41% of P-C4d positive patients had undetectable or low (below the positive cutoff) levels of aPL antibodies at the study visit (Table 2). There are two possible explanations for this apparent lack of correlation between P-C4d and aPL antibodies. First, an extremely low concentration of aPL antibodies, particularly those of the IgM isotype, may not be detectable by conventional immunoassays but may be adequate to bind to platelets and activate/generate C4d in situ on platelet surfaces. Second, it is possible that other platelet-reactive autoantibodies may participate in generating P-C4d in vivo. The latter possibility is supported by the observation that plasma of aPL-negative SLE patients was capable of inducing P-C4d in vitro (Table 5), and that removal of platelet-reactive antibodies by preabsorption of SLE plasma with platelets abolished the in vitro P-C4d-inducing activity (Figure 3B). Several platelet-reactive autoantibodies recognizing normally expressed platelet surface molecules such as glycoprotein IIb/IIIa (GPIIb/IIIa), have previously been reported [47,48]. It is also possible that SLE patients may develop autoantibodies against disease-related "neo" antigens specifically expressed on their platelets. We postulate that such autoantigens/autoantibodies may be present either generally in all SLE patients, or differentially among individual SLE patients. Indeed, the in vitro P-C4d induction experiments have shown that plasma of a group of SLE patients was capable of inducing C4d deposition on platelets derived from both healthy individuals and SLE patients (Table 5; SLE patients #1 and #3), whereas plasma from another group of SLE patients was only able to induce C4d deposition on platelets from a few selected SLE patients (Table 5; SLE patients #2 and #4). These results suggest that the former group of SLE patients may have developed autoantibodies that react with antigens normally expressed on platelets, whereas the latter group of SLE patients may have developed autoantibodies that recognize "neo" antigens expressed only on SLE platelets. Further studies are warranted to identify such potential platelet-reactive autoantibodies in SLE patients.

In summary, the present study not only advances previous findings regarding the potential role of aPL antibodies in generating P-C4d in SLE patients, but, together with our recent study on T-cell-bound C4d [46], also solidifies further the mechanistic model of autoantibody-mediated generation of CB-CAP signatures in SLE patients.

Acknowledgments: The authors would like to acknowledge our colleagues in the Lupus Center of Excellence, Autoimmunity Institute, Allegheny Health Network for providing patient blood samples for this study. We also thank Nicole Wilson, Susan Hebda, and Glennys Smith for coordinating the study. This work was supported by a grant from the Department of Defense (Peer Reviewed Medical Research Grant W81XWH-06-2-0038).

Author Contributions: Chau-Ching Liu and Joseph Ahearn designed the study. Chau-Ching Liu and Travis Schofield performed the experiments. Chau-Ching Liu and Amy Tang analyzed and interpreted the data. Susan Manzi recruited and examined patients, and provided clinical information. Chau-Ching Liu and Joseph Ahearn prepared the manuscript. Susan Manzi and Amy Tang critically reviewed the manuscript.

Conflicts of Interest: Joseph Ahearn, Susan Manzi, and Chau-Ching Liu are the inventors of the CB-CAPs lupus biomarker technology, which is owned by the University of Pittsburgh. SM and JA are consultants to Exagen Diagnostics, Inc., which has an exclusive license to the CB-CAPs technology.

References

1. Manzi, S.; Meilahn, E.N.; Rairie, J.E.; Conte, C.G.; Medsger, T.A., Jr.; Jansen-McWilliams, L.; D'Agostino, R.B.; Kuller, L.H. Age-specific incidence rates of myocardial infarction and angina in women with systemic lupus erythematosus: Comparison with the framingham study. *Am. J. Epidemiol.* **1997**, *145*, 408–415. [CrossRef] [PubMed]

2. Ward, M.M. Premature morbidity from cardiovascular and cerebrovascular diseases in women with systemic lupus erythematosus. *Arthritis Rheum.* **1999**, *42*, 338–346. [CrossRef]

3. Nikpour, M.; Urowitz, M.B.; Gladman, D.D. Premature atherosclerosis in systemic lupus erythematosus. *Rheum. Dis. Clin. N. Am.* **2005**, *31*, 329–354. [CrossRef] [PubMed]

4. Haque, S.; Bruce, I.N. Therapy insight: Systemic lupus erythematosus as a risk factor for cardiovascular disease. *Nat. Clin. Pract. Cardiovasc. Med.* **2005**, *2*, 423–430. [CrossRef] [PubMed]
5. Yaniv, G.; Twig, G.; Shor, D.B.; Furer, A.; Sherer, Y.; Mozes, O.; Komisar, O.; Slonimsky, E.; Klang, E.; Lotan, E.; et al. A volcanic explosion of autoantibodies in systemic lupus erythematosus: A diversity of 180 different antibodies found in sle patients. *Autoimmun. Rev.* **2015**, *14*, 75–79. [CrossRef] [PubMed]
6. McNeil, H.P.; Simpson, R.J.; Chesterman, C.N.; Krilis, S.A. Anti-phospholipid antibodies are directed against a complex antigen that includes a lipid-binding inhibitor of coagulation: β2-glycoprotein I (apolipoprotein H). *Proc. Natl. Acad. Sci. USA* **1990**, *87*, 4120–4124. [CrossRef] [PubMed]
7. Bevers, E.M.; Galli, M. β2-glycoprotein I for binding of anticardiolipin antibodies to cardiolipin. *Lancet* **1990**, *336*, 952–953. [CrossRef]
8. Galli, M.; Comfurius, P.; Maassen, C.; Hemker, H.C.; de Baets, M.H.; van Breda-Vriesman, P.J.; Barbui, T.; Zwaal, R.F.; Bevers, E.M. Anticardiolipin antibodies (ACA) directed not to cardiolipin but to a plasma protein cofactor. *Lancet* **1990**, *335*, 1544–1547. [CrossRef]
9. Bevers, E.M.; Galli, M.; Barbui, T.; Comfurius, P.; Zwaal, R.F. Lupus anticoagulant igg's (LA) are not directed to phospholipids only, but to a complex of lipid-bound human prothrombin. *Thromb. Haemost.* **1991**, *66*, 629–632. [PubMed]
10. Bertolaccini, M.L.; Hughes, G.R.; Khamashta, M.A. Revisiting antiphospholipid antibodies: From targeting phospholipids to phospholipid binding proteins. *Clin. Lab.* **2004**, *50*, 653–665. [PubMed]
11. Petri, M. Epidemiology of the antiphospholipid antibody syndrome. *J. Autoimmun.* **2000**, *15*, 145–151. [CrossRef] [PubMed]
12. Horbach, D.A.; van Oort, E.; Donders, R.C.; Derksen, R.H.; de Groot, P.G. Lupus anticoagulant is the strongest risk factor for both venous and arterial thrombosis in patients with systemic lupus erythematosus. Comparison between different assays for the detection of antiphospholipid antibodies. *Thromb. Haemost.* **1996**, *76*, 916–924. [PubMed]
13. Greaves, M. Antiphospholipid antibodies and thrombosis. *Lancet* **1999**, *353*, 1348–1353. [CrossRef]
14. Galli, M.; Luciani, D.; Bertolini, G.; Barbui, T. Lupus anticoagulants are stronger risk factors for thrombosis than anticardiolipin antibodies in the antiphospholipid syndrome: A systematic review of the literature. *Blood* **2003**, *101*, 1827–1832. [CrossRef] [PubMed]
15. Nojima, J.; Kuratsune, H.; Suehisa, E.; Kitani, T.; Iwatani, Y.; Kanakura, Y. Strong correlation between the prevalence of cerebral infarction and the presence of anti-cardiolipin/β2-glycoprotein I and anti-phosphatidylserine/prothrombin antibodies—Co-existence of these antibodies enhances ADP-induced platelet activation in vitro. *Thromb. Haemost.* **2004**, *91*, 967–976. [PubMed]
16. Lim, W. Thrombotic risk in the antiphospholipid syndrome. *Semin. Thromb. Hemost.* **2014**, *40*, 741–746. [CrossRef] [PubMed]
17. Sorice, M.; Longo, A.; Capozzi, A.; Garofalo, T.; Misasi, R.; Alessandri, C.; Conti, F.; Buttari, B.; Rigano, R.; Ortona, E.; et al. Anti-β2-glycoprotein I antibodies induce monocyte release of tumor necrosis factor alpha and tissue factor by signal transduction pathways involving lipid rafts. *Arthritis Rheum.* **2007**, *56*, 2687–2697. [CrossRef] [PubMed]
18. Salmon, J.E.; de Groot, P.G. Pathogenic role of antiphospholipid antibodies. *Lupus* **2008**, *17*, 405–411. [CrossRef] [PubMed]
19. Rand, J.H.; Wu, X.X.; Quinn, A.S.; Taatjes, D.J. The annexin A5-mediated pathogenic mechanism in the antiphospholipid syndrome: Role in pregnancy losses and thrombosis. *Lupus* **2010**, *19*, 460–469. [CrossRef] [PubMed]
20. Satta, N.; Kruithof, E.K.; Fickentscher, C.; Dunoyer-Geindre, S.; Boehlen, F.; Reber, G.; Burger, D.; de Moerloose, P. Toll-like receptor 2 mediates the activation of human monocytes and endothelial cells by antiphospholipid antibodies. *Blood* **2011**, *117*, 5523–5531. [CrossRef] [PubMed]
21. Perez-Sanchez, C.; Ruiz-Limon, P.; Aguirre, M.A.; Bertolaccini, M.L.; Khamashta, M.A.; Rodriguez-Ariza, A.; Segui, P.; Collantes-Estevez, E.; Barbarroja, N.; Khraiwesh, H.; et al. Mitochondrial dysfunction in antiphospholipid syndrome: Implications in the pathogenesis of the disease and effects of coenzyme Q_{10} treatment. *Blood* **2012**, *119*, 5859–5870. [CrossRef] [PubMed]
22. Giannakopoulos, B.; Krilis, S.A. The pathogenesis of the antiphospholipid syndrome. *N. Engl. J. Med.* **2013**, *368*, 1033–1044. [CrossRef] [PubMed]

23. Martinuzzo, M.E.; Maclouf, J.; Carreras, L.O.; Levy-Toledano, S. Antiphospholipid antibodies enhance thrombin-induced platelet activation and thromboxane formation. *Thromb. Haemost.* **1993**, *70*, 667–671. [PubMed]

24. Galli, M.; Bevers, E.M.; Comfurius, P.; Barbui, T.; Zwaal, R.F. Effect of antiphospholipid antibodies on procoagulant activity of activated platelets and platelet-derived microvesicles. *Br. J. Haematol.* **1993**, *83*, 466–472. [CrossRef] [PubMed]

25. Nojima, J.; Suehisa, E.; Kuratsune, H.; Machii, T.; Koike, T.; Kitani, T.; Kanakura, Y.; Amino, N. Platelet activation induced by combined effects of anticardiolipin and lupus anticoagulant IGg antibodies in patients with systemic lupus erythematosus—Possible association with thrombotic and thrombocytopenic complications. *Thromb. Haemost.* **1999**, *81*, 436–441. [PubMed]

26. Shi, T.; Giannakopoulos, B.; Yan, X.; Yu, P.; Berndt, M.C.; Andrews, R.K.; Rivera, J.; Iverson, G.M.; Cockerill, K.A.; Linnik, M.D.; et al. Anti-β2-glycoprotein I antibodies in complex with β2-glycoprotein I can activate platelets in a dysregulated manner via glycoprotein Ib-IX-V. *Arthritis Rheum.* **2006**, *54*, 2558–2567. [CrossRef] [PubMed]

27. Urbanus, R.T.; Pennings, M.T.; Derksen, R.H.; de Groot, P.G. Platelet activation by dimeric β2-glycoprotein I requires signaling via both glycoprotein Ibα and apolipoprotein E receptor 2'. *J. Thromb. Haemost.* **2008**, *6*, 1405–1412. [CrossRef] [PubMed]

28. Arad, A.; Proulle, V.; Furie, R.A.; Furie, B.C.; Furie, B. β2-glycoprotein-1 autoantibodies from patients with antiphospholipid syndrome are sufficient to potentiate arterial thrombus formation in a mouse model. *Blood* **2011**, *117*, 3453–3459. [CrossRef] [PubMed]

29. Meroni, P.L.; Borghi, M.O.; Raschi, E.; Tedesco, F. Pathogenesis of antiphospholipid syndrome: Understanding the antibodies. *Nat. Rev. Rheumatol.* **2011**, *7*, 330–339. [CrossRef] [PubMed]

30. Holers, V.M.; Girardi, G.; Mo, L.; Guthridge, J.M.; Molina, H.; Pierangeli, S.S.; Espinola, R.; Xiaowei, L.E.; Mao, D.; Vialpando, C.G.; et al. Complement C3 activation is required for antiphospholipid antibody-induced fetal loss. *J. Exp. Med.* **2002**, *195*, 211–220. [CrossRef] [PubMed]

31. Girardi, G.; Redecha, P.; Salmon, J.E. Heparin prevents antiphospholipid antibody-induced fetal loss by inhibiting complement activation. *Nat. Med.* **2004**, *10*, 1222–1226. [CrossRef] [PubMed]

32. Fischetti, F.; Durigutto, P.; Pellis, V.; Debeus, A.; Macor, P.; Bulla, R.; Bossi, F.; Ziller, F.; Sblattero, D.; Meroni, P.; et al. Thrombus formation induced by antibodies to β2-glycoprotein i is complement dependent and requires a priming factor. *Blood* **2005**, *106*, 2340–2346. [CrossRef] [PubMed]

33. Davis, W.D.; Brey, R.L. Antiphospholipid antibodies and complement activation in patients with cerebral ischemia. *Clin. Exp. Rheumatol.* **1992**, *10*, 455–460. [PubMed]

34. Navratil, J.S.; Manzi, S.; Kao, A.H.; Krishnaswami, S.; Liu, C.C.; Ruffing, M.J.; Shaw, P.S.; Nilson, A.C.; Dryden, E.R.; Johnson, J.J.; et al. Platelet C4d is highly specific for systemic lupus erythematosus. *Arthritis Rheum.* **2006**, *54*, 670–674. [CrossRef] [PubMed]

35. Kao, A.H.; McBurney, C.A.; Sattar, A.; Lertratanakul, A.; Wilson, N.L.; Rutman, S.; Paul, B.; Navratil, J.S.; Scioscia, A.; Ahearn, J.M.; et al. Relation of platelet c4d with all-cause mortality and ischemic stroke in patients with systemic lupus erythematosus. *Transl. Stroke Res.* **2014**, *5*, 510–518. [CrossRef] [PubMed]

36. Peerschke, E.I.; Yin, W.; Alpert, D.R.; Roubey, R.A.; Salmon, J.E.; Ghebrehiwet, B. Serum complement activation on heterologous platelets is associated with arterial thrombosis in patients with systemic lupus erythematosus and antiphospholipid antibodies. *Lupus* **2009**, *18*, 530–538. [CrossRef] [PubMed]

37. Lood, C.; Eriksson, S.; Gullstrand, B.; Jonsen, A.; Sturfelt, G.; Truedsson, L.; Bengtsson, A.A. Increased c1q, c4 and c3 deposition on platelets in patients with systemic lupus erythematosus—A possible link to venous thrombosis? *Lupus* **2012**, *21*, 1423–1432. [CrossRef] [PubMed]

38. Lood, C.; Tyden, H.; Gullstrand, B.; Sturfelt, G.; Jonsen, A.; Truedsson, L.; Bengtsson, A.A. Platelet activation and anti-phospholipid antibodies collaborate in the activation of the complement system on platelets in systemic lupus erythematosus. *PLoS ONE* **2014**, *9*, e99386. [CrossRef] [PubMed]

39. Mehta, N.; Uchino, K.; Fakhran, S.; Sattar, M.A.; Branstetter, B.F.T.; Au, K.; Navratil, J.S.; Paul, B.; Lee, M.; Gallagher, K.M.; et al. Platelet C4d is associated with acute ischemic stroke and stroke severity. *Stroke* **2008**, *39*, 3236–3241. [CrossRef] [PubMed]

40. Tan, E.M.; Cohen, A.S.; Fries, J.F.; Masi, A.T.; McShane, D.J.; Rothfield, N.F.; Schaller, J.G.; Talal, N.; Winchester, R.J. The 1982 revised criteria for the classification of systemic lupus erythematosus. *Arthritis Rheum.* **1982**, *25*, 1271–1277. [CrossRef] [PubMed]

41. Hochberg, M.C. Updating the american college of rheumatology revised criteria for the classification of systemic lupus erythematosus. *Arthritis Rheum.* **1997**, *40*, 1725. [CrossRef] [PubMed]

42. Burchett, W.W.; Ellis, A.R.; Harrar, S.W.; Bathke, A.C. Nonparametric inference of multivariate data: The R package npmv. *J. Stat. Softw.* **2017**, *76*, 18. [CrossRef]

43. Machin, S.J. Platelets and antiphospholipid antibodies. *Lupus* **1996**, *5*, 386–387. [CrossRef] [PubMed]

44. Liu, C.C.; Manzi, S.; Kao, A.H.; Navratil, J.S.; Ahearn, J.M. Cell-bound complement biomarkers for systemic lupus erythematosus: From benchtop to bedside. *Rheum. Dis. Clin. N. Am.* **2010**, *36*, 161–172. [CrossRef] [PubMed]

45. Ahearn, J.M.; Manzi, S.; Liu, C.C. The lupus biomarker odyssey: One experience. *Methods Mol. Biol.* **2014**, *1134*, 17–35. [PubMed]

46. Liu, C.C.; Manzi, S.; Ahearn, J.M. Antilymphocyte autoantibodies generate T cell-C4d signatures in systemic lupus erythematosus. *Transl. Res.* **2014**, *164*, 496–507. [CrossRef] [PubMed]

47. Fujisawa, K.; Tani, P.; O'Toole, T.E.; Ginsberg, M.H.; McMillan, R. Different specificities of platelet-associated and plasma autoantibodies to platelet GPIIb-IIIa in patients with chronic immune thrombocytopenic purpura. *Blood* **1992**, *79*, 1441–1446. [PubMed]

48. Macchi, L.; Rispal, P.; Clofent-Sanchez, G.; Pellegrin, J.L.; Nurden, P.; Leng, B.; Nurden, A.T. Anti-platelet antibodies in patients with systemic lupus erythematosus and the primary antiphospholipid antibody syndrome: Their relationship with the observed thrombocytopenia. *Br. J. Haematol.* **1997**, *98*, 336–341. [CrossRef] [PubMed]

MDPI

St. Alban-Anlage 66

4052 Basel, Switzerland

Tel. +41 61 683 77 34

Fax +41 61 302 89 18

http://www.mdpi.com

Antibodies Editorial Office

E-mail: antibodies@mdpi.com

http://www.mdpi.com/journal/antibodies

www.ingramcontent.com/pod-product-compliance
Lightning Source LLC
Chambersburg PA
CBHW051910210326
41597CB00033B/6100